Brinton Webb Woodward

Old Wine in New Bottles

Brinton Webb Woodward

Old Wine in New Bottles

ISBN/EAN: 9783337327958

Printed in Europe, USA, Canada, Australia, Japan

Cover: Foto ©berggeist007 / pixelio.de

More available books at **www.hansebooks.com**

OLD WINE
IN NEW BOTTLES.

FOR OLD AND NEW FRIENDS.

BY

BRINTON W. WOODWARD.

"Old Wood to Burn! Old Wine to Drink!
Old Friends to Trust! Old Authors to Read!"

TO
"THE OLD AND NEW CLUB,"
OF LAWRENCE,
AT WHOSE INSTANCE MANY OF THESE PAPERS WERE FIRST WRITTEN

THIS VOLUME IS FRATERNALLY DEDICATED.

INTRODUCTORY.

ONCE IN A WHILE comes a superior season wherein the vintage is again a rich one: once in a great while a book is projected into the world, full of grand, new, inspiring thoughts.

This book is not one of the few. It will simply find its place, if any, among the countless thousands that take some of the old ideas already in the world, and give them a form somewhat new.

With the mass of us, "there is nothing new under the sun." A thought comes to us, and we clothe it in words; but it had already been masquerading up and down through the world of literature for ages,—and before literature began it had been in the minds of men that built pyramids. (Somebody has, no doubt, made this identical observation before.) To the expression indeed, we may give some individuality of form; the idea has belonged to the race.

In the dialectics of theology and metaphysics we thresh over the same old straw. In literature we fine, and decant, and bottle up the old wine. We pour over the old liquor into new packages, and put on labels of our own. Haply we filter away the lees and dregs which time had precipitated to the bottom.

In setting down herein some observations on people and places, and pictures and books,—the writer is by no means presumptuous enough to imagine that he is adding an iota to the sum of human knowledge! He is simply taking some of the old wine already in stock, and decanting it into his own bottles. Of course he has made his own selection of vintages,—and takes occasion to express some opinion of the quality, or to call attention to the *bouquet*. It will now be entirely in order for any other vintner to declare that these particular wines are not esteemed at all by connoisseurs—and in fact have no commercial value: that they never were worth bottling, or that the work has been so slovenly done they had been better left undisturbed. He shall be free to pronounce the wine very thin, indeed,—but not fairly, we trust, that the beverage is uncommon sour!

To come down to plain English—and the author would be particularly glad could he come up to good plain English—here are a number of essays, sketches, reminiscences of travel,—and a few bits of verse. All assorted together and fairly bound into a book. Take it for what it is worth! The type is new, though the ideas may have no originality;—the paper is good, whatsoever the style of expression. Go to! what more would you!

It only remains to add that quite a goodly share of these papers were first published in the columns of the Lawrence JOURNAL, under the pseudonym of THE LOUNGER. Assumed in the first place to veil a personality, the *sobriquet*—albeit rather trite—is here retained, as serving to avoid the too frequent recurrence of the pronoun "of the first part."

CONTENTS.

	PAGE.
INTRODUCTORY	v
HONORE DE BALZAC	1
BALZAC AND THACKERAY	10
A LITERARY FORECAST	16
WAS BURNS COLOR-BLIND TO THE SEA	17
SCOTT-LAND	26
GEORGE FOX AND HIS JOURNAL	33
THE QUAKER WEDDING	42
TWO SCHOOLS OF FICTION	43
TWO TRAVELERS OF SUCH A CAST	60
THE REALIST IN ART	65
FROM REALISM TO IDEALISM	73
THE OLD KNICK	83
PUTNAM'S MONTHLY	89
THE GOLDEN AGE	97
1855 TO 1854, GREETING	98

HONORÉ DE BALZAC.

It was no small or mean ambition that Balzac entertained when he started out to write the works that have made him famous—no slight task that he consciously set before him and to which he devoted his literary life.

It was simply this, as he tells us in an introduction, written long afterward,—to compass the whole round of human experience—to traverse the whole range of human emotions, passions, motives—to trace every spring of action to its source—to map out the whole world of human life.

This was all! At least this was all he put before himself in the outset. Afterward, as he came to a completer knowledge of his own mental and spiritual capacities, he added to the above a solution of the problem of human destiny—and became Metaphysician and Theosophist as well as Novelist Universal.

But in the outset, as I have said, he proposed only to know and record the whole of human life. "La Comédie de la *Vie Humaine" is the comprehensive title then assumed by him to include all his books. This title has a somewhat cynic sound, akin to that of the lines "All the world's a stage and all the men and women merely players"—players in a frivolous comedy—"The Comedy of Human Life," observed, say from behind the scenes, by some Master of the drama, some Shakespeare or Balzac, who knows every tone and trick and gesture of the actors and may put them all, in turn, with all their vaunted airs and graces, into a greater comedy of his own.

Yet Balzac may have meant to use the word "comédie" in a broader sense—that of the Drama of Human Life. Take it in this sense and who shall say that the ambition is not one worthy of the greatest artist in letters that ever lived!

In some respects Balzac was fairly well equipped for his task. His observation of men and facts and places—as well those of the provinces as of Paris itself—was both minute and extensive, his knowledge comprehensive, his faith in himself boundless. Besides this he was master of a style which, viewed through the happy medium of Miss Wormeley's translations, we find eminently clear, direct and forcible.

It is true that the "whole of human life" in his view appears to include only the society of his own country and his own time but, the man who can faithfully picture the life of a single country and one age, has marched a long step toward the representation of all life, for

"Human hearts remain the same,—the sorrow and the sin.
The loves and hopes and fears of old are to our own akin."

But it was not a broad comprehensive survey of the whole of that life in one view, and its reproduction in one book, that Balzac attempted. He was perfectly aware that the totality of life cannot be taken in at one glance—its infinitely varying elements be synthetized into the personalities of one limited set of people—its thousand multifarious experiences be compressed in their record within the limits of one volume.

History even would teach him that the conquerors of the world had proceeded by taking possession of its provinces and kingdoms one after another in due succession, until there were "no more left to conquer." He knew that the whole was the sum of all its parts—and the better way to take in a pie or a cake is to proceed a bite or a slice at a time. It is not wise to "do" all the galleries of the Louvre before breakfast. Better take "a day off"—and give the matter full justice!

So Balzac, in effect says to us: It is true that it is the geography of the whole world that I am going to give you—but behold, I have divided it into a set of dissected maps of the different countries. Now we will proceed, seriatim, to take each one apart and afterward reconstruct it—then, finally, when you have had the whole in turn, *voila!* you have had the whole of human life—say in a detached series of some forty of my books!

For example, here is the kingdom of Avarice—quite a large kingdom this and one that has always maintained a leading part in the affairs of mankind! We will dissect that in the person and actions of Monsieur Grandet! When you come to comprehend all the thousand little mean motives and acts of Monsieur Grandet, the house in which he lived, the servant who toiled for him all her life long, the family which too served and suffered, the people of his surrounding who bowed down to him and plotted year by year for his daughter and heiress—when you come to see all these clearly you will have a good understanding of *Avarice*—once for all.

Then there is Speculation—that is a province that extends its borders in a good many directions. Take my Cæsar Birotteau! This will show you a little different department of human life and we will trace this passion and its unfortunate exemplifications in the history of a good, honest man, in the bourgeois life of Paris, rather than in that of the provincial town. Note with me, how often a weak-minded man manages to get along in the world successfully and acquire a reputation for shrewdness and force of character just through favoring circumstances, and attending to a single line of business which he understands—and then how quickly he flies all to pieces when a lot of scheming swindlers get hold of him and inflame him with this fever of speculation! Yes, after you read my Cæsar Birotteau you will know that side of human life pretty well—and you will never be likely to be drawn into town lot speculations around the Madeleine—nor to put your money into Oklahoma nor Southern California after the boom is bursted—though you may go in the very next time when the boom is on— say at Tacoma, Seattle or Spokane Falls.

Then if you wish to see what seems a still more seamy side of life, read my masterpiece, Père Goriot. This illustrates the passion of paternal love carried to fatuity. In the gallery of representative fools that I, Balzac, am accumulating for your edification, let us place Père Goriot as one of the most conspicuous, yet one who possibly takes hold of your sympathies when he should rather excite your reprehension and contempt. Poor Père Goriot! Is it the moral of thy fate, that *he* is the supreme fool in life who allows the primitive passions and affections which we inherit from our animal ancestors, to possess and control him to the absolute abdication of all his reasoning faculties; who allows the blind, elemental passion of paternity to obscure every other sentiment, till it usurps absolutely the place of conscience and turns decency out of doors! Who wishes to be a Père Goriot, and meet his inevitable fate—yet how many go halfway toward apotheosizing parental love under the name of duty!

Now whether this method of Balzac affords exercise

of the proper solution of all the problems of life, is, perhaps, open to question. It scarce pretends indeed to reach the centers of things by any one great master-stroke, by any one meridian section of cleavage, laying bare the vital nerves and veins and tissues—it displays not like Browning some "Pomegranate,"

> "Which if cut deep down the middle
> Shows a heart within, blood-tinctured
> Of a veined humanity."

It rather takes up humanity piece-meal and treats it topically. Is it therefore a proper or a practical method in any degree? It has certainly the merit of simplicity, and possibly, in consideration of the extreme complexity of Life's machine, the only way is to take the clock to pieces and lay it on the table—just as the medical professors and demonstrators lay the anatomic "subject" on the table, and proceed to show you all the pieces and instruct you what is the separate function of each.

And when you understand the machine—or the human body—by parts, and you see the master mechanic—or the surgeon--place a few of them together again, you will know all about the functions and life processes! Possibly you will be able to put the clock together again and make it go! The man—or what is left of that man —you can't put together, or make him go, for the "go" has all gone out of him—but, perhaps, understanding all these things, you will be able to make the living man go just as you wish—if you comprehend all of Balzac— especially his metaphysics, and his esoteric Theosophy; and have added to that all the Psychology and the metaphysical Christian Science of to-day; something which Balzac seems to have forecasted sixty years ago.

\# x * * * *

At all events there is a certain grandeur of simplicity about the literary method and manner of Balzac which seems to justify to some extent the application of the word Shakespearian. Not only Frenchmen—who are popularly supposed not to understand Shakespeare at all —but certain English and American critics have applied this term to Balzac. Were it not for this authority, the writer would fear that in estimating the method of Balzac with the method of Shakespeare at his ordinary level, he was simply betraying his lack of thorough comprehension of either. It has seemed to him that in their manner of producing an effect—the intended effect—upon the mind of the reader, the vivid, the intensified, the deepened impression of a character that stands for one thing through all—there was a similitude.

Your Grandet, for example, is presented as a miser, pure and simple, unmitigated and unrelieved by any lapses into generosity or ordinary social feeling,—just as your Richard the Third is put before you, an unmitigated butcher and brute. The man is placed upon the canvass with a few sweeps of a coarse brush daubed to the handle with heavy color. You get the intended effect at once and unmistakably. Père Goriot is a doting, fatuous old father, who gives over his fortune, and subsequently melts down his silver spoons and devotes himself to cold and starvation, to feed the vices of his daughters—with no more glimmering of common sense than King Lear exhibits in giving up his kingdom to those harpies, Regan and Goneril.

Neither writer leaves you in any incertitude as to what his characters will do in any case. Given the people as pictured forth, and they will act according to the rule of automatons—they will "fulfil their destiny." The "bears and lions" will always "growl and fight, for 'tis their nature to." This is the rule of "types" and why though easy to comprehend in the first instance, they are apt to become a trifle uninteresting. This is the weakness of fiction constructed upon that plan. When once you have been given the knowledge of their being, the secret of their springs, with the key in the hand of your imagination, you can wind them up and they will go of themselves, almost as well as if their author was manipulating them. It is like looking at Mrs. Jarley's wax figures. Richard the Third will continue to exclaim, as if through sheer force of habit, "off with his head"—and Shylock will never intermit his demand for sixteen ounces of raw left lung!

By the way, the essayist must confess that even as a child reader, he became so heartily sick of this damnable iteration of Shylock, that he would have gone to any length to shut it off, and hence was almost ready to justify that thinnest and most contemptible of quibbles and subterfuges employed by the imported, bogus umpire who ruled Shylock "caught out on a foul" before he had made his first base.

But while the chief representative characters of Balzac are more or less affected by the stiffness and weakness inseparable from "types," most of his other creations are clearly outlined, well rounded and vigorous, life-like personalities. Strongly individualized, for instance, are his examples of the "bourgeoisie," whom Balzac always delights to render. In his transcription of these, our author is eminently happy and successful. Their shops, their homes, their occupations and habits, their sordid toils and their vulgar enjoyments, their honest virtues and their petty weaknesses—what other French writer ever knew, or knowing, ever told them half so well!

It is with this class especially that Balzac is most at home, and it is within its ranks, with all their absurdities and little vanities, with all their limitations of ignorance and narrowness, that he discovers most of the saving virtue that exists in French society: the men are honest and faithful to their engagements, the women are sensible home loving and pure: all are alike industrious and thrifty, while parent and child are mutually joined together in the bonds of self-sacrificing family affection.

When he comes to deal with the plutocracy and aristocracy of his time, the era of the "Citizen King," Louis Philippe, he finds all society honey-combed with dishonesty and reeking with corruption. Is it indeed a true picture of this society that, as the boldest and baldest piece of realism, fits out every married woman in it with a paramour? The writer of Père Goriot would seem to consider indeed no aristocratic household as complete without one, and we are left to infer, easily and plainly, that it is only the youngest of girls, or the woman of the bourgeois class, to whom remains any conception of innocence or virtue.

We have already alluded to the clearness and simplicity of style of Balzac. This is accompanied with a directness and vigor of movement which constitutes a charm in the reading, in these days when we have either so much of the needless elaboration and tedious refining of small things of Realism or the fearfully involved plot and the turgid or hysterical style of composition of Sensationalism in romance. Both the Realistic and Romantic schools of to-day appear to claim Balzac, but truly he belongs absolutely to neither, while possessing some of the most agreeable traits of each. You can always tell what Balzac is driving at—the story goes straight along toward its consummation, which is sometimes, though not always, indicated from the start—there is little involving and no intricacies of plot, no marching and countermarching—the characters proceed in natural order of development—they "tend strictly to business and dont go fooling around."

"Great Homer sometimes nods" or *we* nod as he recounts his tedious list of ships : Shakespeare stops the action of his drama at times while his fellows fire off their big, bombastic speeches : Dickens digresses inconsequently to display the idiosyncracies and oddities of his characters : Thackeray pauses to moralize with genial or cynic observation or to belabor and abuse his own puppets for the very qualities he has imparted to them—then turns on again the music of his measure while the merry dance of the marionettes is happily resumed with the story. But, Balzac—like Tennyson's Brook, pauses not but "goes on forever"—until the story is over and the book is done.

BALZAC AND THACKERAY.

SOME controversy has always cropped out from time to time, respecting the morality of Balzac's writings. It seems somewhat incongruous that one set of people, including indeed some clergymen, should discover in him a moralist of the severest type, while another deplores a writer so immoral that he stopped at no verge of common decency.

Possibly there may be some ground for both opinions. As in the case of the storied viewers of the chameleon in its different conditions, the variable writer may be either a lovely green—or black as jet, just as you happen to come upon him! Taking him thus in his contrasting aspects, Balzac is found both moral and immoral, but, in his most normal condition, I should say wholly unmoral.

In the task to which he addressed himself—to tell us the story of human society—sufficient was it to him to "adorn a tale" without bothering himself to "point a moral." Undoubtedly he conceived that his office was that of the historian, rather than that of the moralist;—his function more truly that of the painter than the preacher. It was his business to reveal to us the whole of human life—the bad equally with the good and the indifferent—and he no more hesitated to treat any aspect of it whatever, on the ground of delicacy or morality, than the eminent, skilled physician or surgeon hesitates to treat any case of disease or deformity that offers itself, or to exemplify it in his clinics.

If in the course of his faithful record, these truthful, scientific, clinic notes, we find that sin has drawn in its train its own inevitable retribution—let the *reader* deduce his moral!

If again, however, weakness of character has brought about equal suffering, misfortune and ruin, it is not Balzac's business to controvert the facts of sociology or to provide that weak-minded goodness alone shall ensure success and happiness in life!

There is a certain philosophy, which—having given society to you as he found it—he leaves you to deduce, but morality is not a matter which concerns him at all; —no more than it did Shakespeare himself, who in Portia discoursing on the divine quality of mercy, or in Hamlet soliloquizing on the transition from the Here to the Hereafter—was not a whit more personally sympathetic than in gross Falstaff with his amours, in jealous Moor, or in malignant Iago! Such indeed is the "impersonal" theory of genius universal, that sees everything clearly, discriminates dispassionately and shares alike and equally in the inmost nature of all;— and, distinguishing with such marvellous insight, partaking thus universally, is by far too large-minded to be anything much less than pantheistic. "Born to a universe," 'tis not for Genius to "narrow his mind" within the range of any one set of sympathies or to limit his soul with any system of morals—

> "Who sees with equal eye, as lord of all,
> A hero perish—or a sparrow fall,
> Atoms or systems into ruin hurled
> And now a bubble burst—and now a world!"

Now, if the writer may be permitted, just let him remark that the "bubble" which should be burst, in his humble opinion, is one that Shakesperiolaters have inflated with gas something like the foregoing, forgetting that no one human mind can be greater than the sum of all other minds in the universe—or any soul, however great, own a prescriptive "right and title" of primogeniture that should exempt and absolve it from operation and observance of moral laws governing in this world of ours.

* * * * * *

Far removed from all this is Thackeray, who never attained, who never could have attained to this sublime height and supreme state of even, dispassionate feeling toward all the facts of life and all the actors in it!

What association is there between the two great names which flutter at the masthead of this paper? Very little, possibly, except that each, perhaps may be regarded as the ablest representative in fiction of his respective nation during the last literary generation, and that both treated especially the society of their time, though Thackeray by no means confined himself to his own age, as witness his "Esmond" and his "Virginians."

But in their literary style they were as far apart as they were in their mode of representation, the picturing of that society. Balzac's style, as we have endeavored to illustrate, is both simple and direct. He tells his story in a straightforward fashion, without expansion or elaboration, though with all due attention to necessary detail to place the scene and the personages properly before you. Having accomplished this necessary outlining, Balzac paints with a few broad strokes, though each is masterly.

Thackeray too has the quality of simplicity, but it is the apparent simplicity of high finish and of great art.

With infinite pains, stroke after stroke, and often intermitting to contemplate his work, the artist proceeds, modifying his broader effects with a lighter touch of color, here a little and there a little—"line upon line"— yea, and "precept upon precept," if such are ever put upon canvas—he goes on, and still the picture grows before your eyes, a marvel of representation, a literary masterpiece. When once finished, with its high lights thrown in and all its half-tones and shadows set in with due *chiaro-oscuro*, the portrait is unmistakable, the characterization is complete. For good or for ill it is done, and you will love or hate it ever after!

It is perhaps the very power and strength of Thackeray, that might of satire and keenness of irony, which have prevented his beauty and tenderness from being as much appreciated as they deserve by the mass of readers, many of whom have been deterred entirely from his reading by the impression that he was a cynic who saw too clearly all the ills and shams of society to discern any good in it. But his cynic tone of satire implies always that there is a soul behind the pen, that is full of righteous indignation against the false pretence, while sympathizing ardently with the genuine and the true.

Say that he looked too deeply below the surface, and saw often in our vaunted deeds of charity even, the vanity and the selfishness that we fain would not suspect ourselves, least of all betray to others; say that the very children whose society in real life he so thoroughly enjoyed, are in his fiction seldom childlike; say that his young men are too prone to be cubs or cads and his old ones, snobs or scoundrels; say that almost the only woman he ever endowed with brains turns out to be an adventuress, while on the good types, he can bestow a soft loving heart only at the expense of inevitable accompaniment with soft, silly head; say that his Pendennis is too often prolix and a prig and his Bayham a bore; but remember that in Ethel Newcome with all her pride, he has pictured a girl full of noble feeling and sensibility, of spirit and of intellect: and forget not that he has given back to the world, as we trust he may have found it in real life and society, the prototype of the perfect gentleman—one who, with all his weaknesses (which render him all the more natural and life-like) stands outlined to us in noble representation, the tenderest and truest, the purest and manliest—brave old Colonel Thomas Newcome! That, too, in picturing the passing away from earth of this example of a noble soul patiently enduring poverty and unmerited disgrace, Thackeray has given to literature a passage so simple, so touching, that it may well endure with the language that he so enriched:—

"At the usual evening hour, the chapel bell began to toll and Thomas Newcome's hands outside the bed feebly beat a time. And just as the last bell struck, a peculiar sweet smile shone over his face, and he lifted up his head a little, and quickly said '*Adsum,*' and fell back. It was the word we used at school when names were called; and lo, he whose heart was as that of a little child, had answered to his name, and stood in the presence of the Master."

Grand old Thackeray! True it is that thou art sometimes too scornful to be agreeable in the reading—but if thou wert perchance too prone to fling the arrows of ridicule and the barbed shafts of satire, it was always at pretension and folly, and falsehood and vice, that they were aimed and therein left transfixed for the world to mark;—it was never honest worth or humble virtue that was scoffed at or derided. If the line of demarcation betwixt our good intent and our selfish promptings was sometimes so sharply drawn that it made us wince with the pain of mortification, we could not say after all that it swerved to the wrong side, albeit it divided too exclusively on that side, more of our own thoughts and acts than we had fain believed!

Grand old Thackeray! Thank fortune that there is nothing universal or impersonal about the quality of thy genius! Thou hadst prepossessions, prejudices, plenty of them and strong ones—against social shams and moral iniquities—and in favor of decency and clean living, manly honor and womanly virtue! In spite of a cynic manner, sometimes assumed it may be to cover and veil a tenderness and delicacy of nature almost to be regarded as effeminate,—whenever a moral conflict is on—always,

"In the strife twixt truth and falsehood,
For the good—or evil side"

thy position is easy enough to be discovered. Masked though it be by the batteries of satire, it is all the more strongly entrenched—and behind them thou carriest on always an effective warfare against the strongholds of dishonor—against the cohorts of vice!

A LITERARY FORECAST.

In his delineation of Balfour of Burley in "Old Mortality," Sir Walter Scott forecasted almost to a prophecy the character of John Brown.

The age, the country and the cause differed widely, but the coincidence in character is striking. It would seem as if the key of every apparent anomaly in the life of the man of Harper's Ferry—anomalies so startling that men even yet refuse to credit some of the well-determined facts of his history—might be found in Scott's portrayal of Balfour.

Devout, but pervert; conscientious, yet unscrupulous and remorseless; bold and direct in purpose and action, yet capable of craft and dissimulation; rash, yet deliberate and calculating; both tender and terrible; a homicide and a hero; a murderer and a martyr;—these are antagonisms that may exist in the life of a religious fanatic, and Sir Walter Scott with his deep insight wrought their seeming incongruities and paradoxes into the type of Balfour—while John Brown exemplified and repeated them in the stress of a great crisis, two centuries after the time to which he legitimately belonged.

WAS BURNS COLOR-BLIND TO THE SEA?

WE had been all day in the Burns country—retracing in one long summer day the footprints of a life journey all too brief, but in the reverse order of its natural course, beginning as we did in the early morning at the Poet's tomb in Dumfries and ending, in the late twilight, at the cottage near Ayr, "where the bard-peasant first drew breath."

Leaving Kirk Alloway, passing a little farther up the slope on the turn to the west—and there before us is the sea! Beyond, dim in the gathering shades of even, are the pale hills of Arran across the Firth of Clyde, and farther still, through an opening in the hills, we faintly descry the Mull of Cantire.

We had been saying to ourselves all day long as we had passed from one beautiful scene to another as fair, or fairer yet, each in its glowing or quiet beauty lending some suggestion of exuberant joyousness, of peaceful calm or pensive melancholy, and all familiar in their every aspect to Robert Burns:—What wonder indeed that he has commemorated so many of them, for who with a heart awake to Nature's charms might not be swayed to some expression of poetic feeling!

Burns has been fitly called Nature's Poet, which reciprocally should make him the poet of Nature. It is true indeed that he comes very close to the heart of nature in many of her tenderest moods and aspects. But after all, is it not a limited nature that he finds and reproduces? It is that of the field, the stream, the grove. He is the Poet of the Plain!

It has lately been told me by one somewhat familiar with the learning of India—the literature of Brahminism—that it would really seem as if that ancient stock of the Aryan race might have been color-blind to one of the finest tints in nature, the blue of the sky, for they have left behind no record of their observation of this beautiful noonday aspect of the heavens, while their literature fairly glows with the gorgeous tints of morn and even.

As we rose to the summit of the hill back of Kirk Alloway, and the beautiful view of the Firth broke upon our eyes, it came to me as a special revelation of wonder that Burns, the royal favorite of Nature, was color-blind to the glory of the sea!

For all the years of his boyhood and into early manhood, he lived within a mile of the sea—its sights before his eye and its sounds often in his ear—and yet, how few hints of this fact are to be found in all his poems!

It may be, as the Firth here lies somewhat sheltered by islands, that the force of marine storms is measurably broken, and that the might and majesty of ocean in its wildest and grandest moods were seldom, if ever, visible to him; but yet there were aspects of it which must have been familiar and which, one would think, would have impressed his imaginative sensibilities; those for instance which Mr. Blaine once recited as longed for and welcomed by the dying Garfield, who, "With wan fevered face tenderly lifted to the cooling breeze looked out wistfully upon the ocean's changing wonders. On its far sails whitening in the morning light; on its restless waves rolling shoreward to break and die beneath the noonday sun; on the red clouds of the evening arching low to the horizon; on the serene and shining pathway of the stars."

Why should not such scenes inspire a poet to noble song, when the man of Politics is moved by them to such eloquent wording—and how should the poet miss all those grand suggestions of Time and Eternity that may be borne with the breakers into the consciousness of Statesman—as intimated in the same eulogy:

"Let us think that his dying eyes read a mystic meaning which only the rapt and parting soul may know! Let us believe that in the silence of the receding world, he heard the great waves breaking on a farther shore and felt already on his wasted brow the breath of the eternal morning."

Thus have the great souls of all times felt and spoken; thus the texture of their language been interwoven with the imagery of the sea!

I may have read imperfectly, but I recall scarce more than a single couplet indicating that Burns was acquainted with the roar of the breakers. Yet this pair of sounding lines, while it negatives thus far the presumption that he was entirely blind and deaf to the beauty and grandeur of the ocean, proves at the same time that the coast of Ayr was far from being unworthy of his poetic notice. This occurs in his "Twa Brigs of Ayr:"

"The tide-swollen Firth with sullen sounding roar
Through the still night dashed hoarse along the shore."

Again, we recall that Burns made more than one visit to the mountainous districts of Scotland—traversing them once as far north as Inverness—but we look through his poems almost in vain for descriptions of the grandeur of mountain scenery. One allusion only I can bring forward which conjoins in one brief stanza, in broad panoramic view, the three magnificent features of Scottish scenery:

> "Here, rivers in the sea were lost!
> There mountains to the sky were tost!
> Here tumbling billows marked the coast
> With surging foam."

Perhaps the very suggestion is supplied in the poem from which I extract this verse—"The Vision"—which may account for Burns' narrowed restriction in the field of nature. Therein he dedicates himself as a poet—not to universal nature by any means, but to that of *his district!* That he was conscious of the grandeur of mountain and of sea, the lines above quoted from his early poems seem about the only evidence. Thenceforth the wide outlook on nature is abandoned, the hills and vales of Ayrshire and Dumfriesshire take its place. Says Wordsworth:

> "Two voices are there: one is of the sea,
> One of the mountains—each a mighty voice."

It seems to me that the truly great poet must give some utterance, some expression, to these two mighty voices! Burns declined to attempt it. Perhaps he was wise and the world has gained through his restraint. He painted no marines or mountain views indeed, but his quiet landscapes are unexcelled in tender beauty and suggestiveness. A sentiment often underlies them beyond the reach of words. More than this—they are not landscapes merely—there are always human figures in the picture, in foreground or the middle distance. He paints man!

After all, I think Burns will live longest as a song writer. In that he did the literature of his native Scotland an immense service. He wedded familiar, sweet and plaintive airs that had long been straying around the country, entirely homeless or sheltering in low-lived quarters, to worthy, pure and sometimes noble words.

The union has become a perpetual one, the songs are sung wherever the language is spoken.

> "And still the burden of his song
> Is love of right—disdain of wrong;
>   Its master chords
> Are Manhood, Freedom, Brotherhood—
> Its discords, but an interlude
>   Between the words.
> * * * * * *
> For now he haunts his native land
> As an immortal youth—his hand
>   Guides every plough—
> He sits beside each ingle-nook,
> His voice is in each rushing brook
>   Each rustling bough."

And if such be the sentiment toward Burns, as voiced by our revered Longfellow, then surely the influence of song is a pervading and mighty one—and there never was a stronger exemplification of the words uttered two hundred years ago by old Andrew Fletcher of Saltoun: "If a man be permitted to make all the ballads, he need not care who should make the laws of a nation."

One thing to be taken into account, in estimating what Burns accomplished, is that he died at the early age of thirty-seven. Many of our best poets have lived to twice this age and performed their best work within the latter half of the lives thus matured. Whether added length of years would have brought any richer life fruits, unless that healthful temperament of body and soul which prolongs life and confers its greatest power had also been bestowed upon Burns is a question hard to answer, as well as the other problem, that of the due measure of responsibility resting on the man himself for the misuse of the good gifts and the indulgence of the hot passions of his nature!

The worst misfortune of all, it seems to me, concerning him, is that so much of the soil that contaminated his life found expression in his verse and has unhappily survived him in the subsequently collected editions. Too late, when broken down by his last sickness, Burns himself regretted that this was likely to be so and that he had not, in time, culled out and destroyed the worthless. It is too much to hope that even then some of the sticks and stones and mud-clods that he had thrown would not have been gathered and bound up into sheaves with the golden grain; but we might have been spared the worst of them. When, oh when, shall we find publishers bold enough to screen out the chaff and dirt from among the wheat and give us clean, expurgated editions of Burns, Byron, Shakespeare?

In the meantime, Burns' earnest pleading may well be applied as a mantle of charity toward himself—his life and work:

> "Then at the balance let's be mute
> We never can adjust it,
> What's done we partly may compute
> But know not what's resisted."

Computing partly by *what's been done*, we can venture to assert, however, that it is perfectly possible for a boy born in poverty and obscurity and nurtured among rude surroundings, to grow up into a great poet—leaving behind him a clean life and noble works of poetic genius, with no unworthy ones—not one line written "which dying he might wish to blot." Some exceptional proofs have been given of this even in the old days, but most striking and noble ones in the poets of our own time and in our own land.

There is an old castle in Scotland—visited, indeed, the day succeeding that of our Burns pilgrimage—whose name is memorable in history. Its tinted walls of red sandstone, overhung with the green of ivy, are picturesque and beautiful even in their decay. At its foot, a lovely stream glides along, rippling over the smooth-worn rocks and pebbles, whose further slopes break down to it in sunny braes, or swell backward into fir-covered knowes and birken-shaws, while on its own side, the wooded glade soon gives place to fair fields and pastures green, leading on to gentle slopes and sunny vales, dotted over with cheerful cottages and adorned with one stately mansion. The landscape, though limited, is one of the "bonniest" in all Scotland!

In the old, fierce days when every man's hand was against his neighbor, the owner of this castle—generous, proud, fiercely independent, acknowledging allegiance to none, of hot passions and bold, reckless tongue—found many foes, chief among which were those in his own unquiet breast. In an unequal contest, where quarter was neither given nor taken, his castle was besieged, its garrison overpowered, its battlements torn down, its strong tower demolished, never to be rebuilt.

Uninhabited it stands and forsaken; but inspiring as well as mournful memories cluster around it and sometimes draw to it the wandering footsteps of the traveller from beyond the seas. In its quiet nooks the field-mouse builds her nest and the wounded hare finds a covert; the daisy and the cowslip sprinkle the neighboring leas; the pleasant river still murmurs by its base, and the song-birds of heaven, the mavis, the merle and the lavrock warble their sweet lays as they build among the branches that yet cast a tender shadow over its time-worn walls.

Historically you may call this Bothwell Castle—but if such could be a poet's emblem, that broken tower I would name--Robert Burns!

In the borders of the Black Forest there is an old

castle whose towers still rise high amid the lofty trees and crown the summit of a mountain which looks abroad over vast stretches of country once called Duchies, Kingdoms, Provinces. On the near hand are dark, rolling forests, breaking off at times against rocky prominences surmounted by other ancient strongholds, whose foundation stones were laid in the days of the Romans.

Beneath nestles a charming green valley, sparkling with bright villas that mark the environs of the most famous watering place in Europe. To the west it looks down upon a magnificent plain, "rich with corn and wine," luxuriant with the life of scattered hamlets, towns and cities—a magnificent panorama spread out for near one hundred miles—the broad valley of the Rhine. Far beyond, the eye reaches, at the horizon, the distant wavy outline of the "Blue Alsatian Mountains." How many elements, both of grandeur and of beauty, doth this fair landscape comprise!

This castle, too, has had its history, its hot youth of ardent contest and defeat; but for the last two hundred years, from its grand seat, it has looked calmly down upon struggles that have convulsed empires.

And yet, even in its decay, it is far from lifeless. Day by day it is sought by young and old alike. On the sunny slope of its terrace, we beheld gay pic-nic groups of pleasure seekers; through its winding courts strays wooing youth with maiden; while the inner hall, though roofless, resounded to our ears with the jocund laugh and happy sports of childhood. The artist, too, comes to fondly reproduce, on paper or canvas, its picturesque features or reminiscences of its old-time glories.

More than this. Across all the grand, old, empty casements, deft hands have strung wires of differing size and tension—"harps that the wandering breezes tune"—and now the old castle knows music sweeter far than was ever awakened by its harpers of old, as the rising or falling winds steal softly or sweep grandly across the chords of these harps of Æolus; now weird and low, and anon swelling in magnificent diapason which reverberates through all its ancient arches, and again subsiding in tender cadence inexpressibly plaintive and touching.

To me, no organ peal resounding amid the dim aisles of vast cathedral, no

> "Bird's clearest carol by fall or by swelling,
> Such magical sense conveys."

as these "airs from heaven" ringing their own melody!

There is naught like it but—the poet of fourscore years singing at eventide, who, as the nearing shades of the dark forest fall upon him, yet looks trustfully out upon the beautiful plain, bright with the beams of the setting sun—and onward to the peaks of the Delectable Mountains bathed in sunset's gold. Still singing because it is in his heart to sing!

Over there they may call this castle the "Alt Schloss Hœhenbaden," but, if this too could be assigned as Poet's emblem, in fancy truer I would name it from loved bard of our own land—Whittier!

Would that the grand name and fame which in future will so justly attach to his, could as worthily rest with thy memory, poor Bobbie Burns!

SCOTT-LAND.

(From a lecture delivered at the State University of Kansas.)

A NEW WORLD—the Realm of Books! Broad continents whose teeming plains or Hesperidean valleys yield the ripened grain or golden fruits of Knowledge;—great seas traversed by a thousand argosies which bear homeward from far-off isles or dim mysterious shores the strange, rich treasures of Imaginative Thought!

Such isles of Romance—such plains of Verity—have been brought within your ken, young student of Kansas University! Such New World you perchance discovered, Columbus-like, and made it your own, when, for the first time, you entered the confines of a room below—The Library!

In the long years before you, it may be your fortune to receive many varied and rich impressions. It were little, indeed, to say that scarce shall one arrive to you that may not remind you of something already traced in yonder volumes.

Traversing to the Atlantic and then across its bounds, you may one day attain to that much desired haven of the American scholar — the Old World of Europe. Standing in certain rooms of the old Tower of London, the Museum Johannes at Dresden, or the Invalides of Paris, you may gather new impressions like the rays of light reflected to your eyes from countless gleaming arms and armor. Like Longfellow at Springfield, you exclaim:

'This is the Arsenal—from floor to ceiling,
Like a vast organ, rise the burnished arms!"

Insensible then, indeed, were you to a thousand historic memories, could you fail to be impressed by the numberless trophies which Time has wrested from the nerveless grasp of mighty warriors of the ages past, from the days of Cœur-de-Lion to Napoleon the Great. "Departed spirits of the mighty dead"—these, their battle-axe and sword and shield, and not the marble mausoleum reared "amid the long drawn aisle and fretted vault,"—these are their fitting monument. Heroes of old, peace to your ashes!

"Your good swords rust,
Your steeds are dust,
Your souls are with the saints, we trust!"

But standing beneath the dome of the British Museum, or in the rooms of the Bodleian at Oxford—noting the vast multitude of *books* piled, range upon range, "from floor to ceiling," while, near by, hundred of cases preserve rare manuscripts that date centuries back of the oldest printed volumes, with autographs of famous kings and still more famous Kings of Letters—you shall scarce fail to be moved to a tenfold greater degree than by any trophies of military greatness, however memorable. These, you exclaim, are the forces that shall influence the world hereafter—the weapons of Knowledge. "*This* is the Arsenal, *these* the burnished arms!"

And yet, if I may gauge your feelings by mine own, I should say that even then your enthusiasm might scarce renew, much less surpass that of your earlier days, when first made free of the University Library. After all, a well selected list of *five thousand volumes comprises the heart of the world's literature, and the first rich zest of the seeking mind is scarce to be transcended.

---
*Now (1890) increased to 11,000 volumes.

As fresh and fair as ever to me is that bright day of boyhood when first I entered a quiet room in an old, brick farm-house in Quaker Pennsylvania and, coming into the possession which a season ticket in its circulating library of 1,600 volumes gave me, went forth from that day into a strange new world—into

"That new world which was the old."

Exploring the wide confines of that world, I one day discovered a new Kingdom—Scott-land! In all the grand realms of literature, the bright sun of genius illumed no land more fair—among all the spells of Romance, to me none more potent than that cast by the mighty magician, the Wizard of the North!

Poesy, History, Romance; wondrous legend, impassioned rhetoric, inspiring thought; these were so wonderfully woven, so inextricably blended, you knew not of the three-fold spell which element was the most powerful nor which held the greatest charm!

He tells of his mythical wizard of the same patronymic, Sir Michael Scott, such was his magic power,

"That when in Salamanca's cave
He list his magic wand to wave,
The bells would ring in Notre Dame."

But what was this compared with the "magic of the mind" exercised by the real Wizard, who, when he lifted his magic wand—the pen—could ring for us the bells of every land from Scotland to Palestine, until we heard the swell of St. Mary's in fair Melrose chiming in unison with those that hung in the minarets of St. Jean D'Acre!

"Bells of the Past, whose long-forgotten music
Still fills the wide expanse,
Tingeing the sober twilight of the Present
With color of romance."

With that magic music ringing in our ears, come trooping forth a wondrous procession, the marvellous yet material inhabitants of Scott-land!

Before us throng the people of differing races, climes and ages, beginning with the Waverley, whose date to the author's day was but "Sixty Years Since," and extending back into the remote, yet not dim perspective of king and swineherd of the Norman Conquest; the hermit-saint, the Red-Cross knight, and turbaned Turk of the days of the Crusades.

In fancy compelled, we follow Richard of the Lion Heart to the plains of Palestine, and with him storm the walls of Acre or hold courteous parley with the princely paynim Saladin! Anon, we are back on English ground, and with gallant Ivanhoe overthrow Bois-Guilbert and Front-de-Bœuf in the tourney lists of Ashby-de-la-Zouche, whilst the fair daughter of Isaac of York looks fearfully on and stately lady Rowena gives the victor's prize! We scale, with young Arthur Phillipson, the crags of Switzerland, to win the smile of sweet Anne of Geierstein, and, at midnight, we are silently lowered with his father, the fearless old Earl of Oxford, into the underground chambers of the dreaded Vehm-Gericht of Germany! With stout Smith-of-the-Wynd we fight along with Clan Quehele and thirst for the extermination of the last intervening champion of Clan Chattan, that we may get one stroke at their recreant chief, the whilom glover's apprentice! We follow the fickle fortunes of Sir Nigel through fierce Alsatia, as we do the wandering steps of bold Quentin Durward through foreign Flanders and fair France! With rugged Balfour of Burley we wield the "sword of the Lord and of Gideon" against the troopers of bloody Claverhouse, at the battle of

Bothwell Brig! Again the sound of revelry is heard, and with the false Earl of Leicester we welcome to Kenilworth Castle, with pomp and rejoicing, "Good Queen Bess," learned and vain, amorous, haughty, and mean, while poor Amy Robsart, the deserted and ill-fated countess, is treacherously dismissed, to mingle her falling tears with the "dews of summer night that fall," at Cumnor Hall. With Julian Avenel and bold Catherine Seyton we conspire to release Mary Stuart from her prison of Lochleven Castle, and the spell of the author over us is as the witchery of the fascinating queen upon the boy page, that whilst her dark guilt is intimated, her misfortunes and her magnetism obscure our better judgment, and we, too, are ready to do and to dare everything to save her from her impending, inevitable fate! Away from lake to the Highlands, in whose fastnesses we lose ourselves with that unique soldier of fortune, Sir Dugald Dalgetty, and vanish with the Children of the Mist, whilst the baying of pursuing bloodhounds is faintly heard in distant defiles below! We dine with Dandie Dinmont, or taste the witch's broth that Meg Merrilies pours scalding into the smuggler's throat, while Dominie Sampson exclaims "Prodigious;" or with faithful, lying, old Caleb Balderstone, we steal the roasted fowls from the spit, for our master's honored guests—a light foil of humor which more completely shades the dark tragedy of Ravenswood and poor Lucy of Lammermoor.

We follow on foot, from "within a mile of Edinboro' Town" all the way to London, the true hearted, the noble though lowly-born Jennie Deans, till we secure from the gracious Queen a pardon for erring sister Effie! We—but it is useless and time fails to recall more of the thousand "beautiful pictures that hang on memory's wall" in the enchanted chambers of that grand old feudal castle, stormed and ruled by the founder of Abbotsford—the Castle of Historic Romance, wherein preserved are all that was best, bravest and most beautiful of all the centuries of Feudalism; for it was to this historic feudal era that belonged the genius of Sir Walter Scott, who lived the century after its departure, but in time to be moulded and swayed by its charm to its perfect revelation:

> "For all his life the charm did talk
> About his path—and hover near,
> With words of promise in his walk,
> And whispered voices at his ear."

And like the enchanted palace to the Fairy Prince, this feudal castle opened to his magic key; from long-century sleep, its voices woke,

> "And buzzed and clacked,
> And all the long-pent stream of life,
> Dashed downward in a cataract."

* * * * * * * * *

And yet, though the kingdom of Scott was wider than the breadth of Europe, there was one little province which, while on its extreme western marge, was yet its center, its heart; and though the least in circumference was by far the greatest of all—little Scotland. Scotland with one "t"!

Macedonia of old, was greater than all the rest of Alexander's world beside, for it bred and inspired Alexander: Rome, than all the rest of the great Roman Empire; little England, than the whole of vast India, Australia and all the islands of the sea. And the land of Sir Walter Scott was, above all else, the little rugged province of his home and love:

> "Land of brown heath and shaggy wood,
> Land of the mountain and the flood."

And so, the genuine lover of Scott has found his dearest imagination ever turning away, from desert of Syria, though traversed by Richard of the Lion Heart and the brave Knight of the Leopard; from peaks of Switzerland; from sunny plains of France; from lawless Whitefriars; from lonely midnight rambles in the Park of Woodstock; from gay joust and tournament on the field of Ashby-de-la-Zouche; from queenly revels even in the halls of Kenilworth; back to that picturesque region of mist and mountain and moorland, where, irradiated with the finest glow of the Author's magic fancy,

> "Every rock and hill and stream
> Appareled with celestial light did seem
> The glory and the freshness of a dream."

GEORGE FOX AND HIS JOURNAL.

UP in the garret of the old Pennsylvania farm-house in which I was born, I made discovery, very early in my boyhood days, of a wonderful book. It was a musty, antiquated volume, that had come down from I know not what ancestor, printed at least a century before, in the old-fashioned type wherein the "s" and "f" are inseparably confused to the modern eye; a bulky and ponderous quarto, about the size originally of "the family Bible that lay on the stand" down stairs. A homely old book, with its heavy, coarse, whitey-brown, uncalendered paper—the front one of its leather covers departed, and with it some of the earlier pages beside. But this ragged old volume, or what was left of it, was a veritable mine of romance to my youthful imagination, as much so as any rare edition of Froissart to the antiquary, or a fragment of Thomas the Rhymer to Sir Walter Scott.

In the early spring or late autumn, when the farm work was apt to be interrupted by storms, I sought the old garret and devoured the pages—reading them many times over I dare say, often oblivious to the passing hours, and turning the leaves at last with fingers chilled, numb and blue. Some forty-odd years have passed since then, but still the vision of the old attic comes freshly back to-day: its bare shingles and rafters overhead, the old spinning-wheel and reel in one corner, a discarded meal chest for a table, and the *wonderful book*, which I seem ever reading in association with that most melodious of musical monotones, the patter of the rain drops on the roof.

Now this remarkable work, for such I considered it, and so it seems to me even yet, was none of those famous, old-time classics of boyhood, "The Arabian Nights," "Robinson Crusoe," "Pilgrim's Progress," nor yet "Don Quixote," nor even the half-historic, half-mythic "Lives" of Plutarch, but merely the autobiography of a man who began life as a shoemaker's apprentice and ended it as a Quaker preacher—the "Journal of George Fox." It was the simple life record of the founder of Quakerism. But to begin with, it had one great, rightful hold on the mind of youth (say what we may as to a child's fondness for fiction), it was all true, or at least the earnest writer firmly believed it to be true, which, I take it, is about the same in effect to a boy of ten. Moreover, it was no tame and tedious recital of ministerial wanderings and ponderings. There was much in it I could not understand, but it was brim full of incident and adventure, bursts of rugged, untaught eloquence, and passages of kindling fire, where fervid rhapsody seemed mounting into inspired sublimity. The writer had traversed all of Great Britain and many other countries beside, and wherever he went had created a sensation, a tumult, a whirlwind; had gone unannounced and uninvited into "steeple houses" (as he termed the churches), and, like Christ in the Temple, had therein boldly denounced the traffickers in religion; had stood in the market-places and highways and reproved the besetting sins of the people; had challenged the authority of Chief Justice of England, and, in person or by message, had sharply catechized or prophetically warned lord protectors. crowned kings and sovereign pontiffs of Rome. He had slept under hedges and refuged in ditches: had been literally floored by the Bible in the hands of infuriated clergymen whom he had just effectually "floored" in discussion; had been stoned and beaten, well nigh murdered by raging populace; dragged to jails by rude soldiery—loathsome jails, whose horrors are happily unimagined in these days—languished therein for months and years, and, when finally released, renewed his life-work as fresh, as unconquerable as ever. Again, he was a seer and a prophet; had wonderful visions, wrought miracles, invoked judgments, and foretold miseries and destructions to come. But time would fail to catalogue a tithe of his doings, or adequately describe his wonderful Journal, whose chief merit, after all, is the key it furnishes as to the inner history of that interesting sect—the Quakers.

What constitutes Quakerism? The ordinary observer of externals merely, might give as incorrect an answer as if he should describe the fruit of the cocoa-palm by its outer covering—something uniformly, homely and dry, hard and husky; yet to those familiar with the real nature, the same shell or outer garb would be suggestive of the cool, refreshing richness within, while the beauteous and luscious odored pomegranate should prove worthless save to scent and sight.

Though these odd and quaint externals may seem absurd to the multitude, they all had, once at least, the significance of vital principles, which the wearers were willing to uphold through persecution unto death, and are, perhaps, the useless, well-nigh obsolete coverings of a religious faith very much like Christianity in its original purity, and as lovely as the world e'er saw!

The rise of the people called Quakers was one of the

greatest anomalies of any age. In the midst of a violent civil war—the first great Revolution of England—when every man's hand was against his neighbor, sprang up this sect, one of whose leading principles was the total abandonment of brute force and the substitution of peace and good-will for the sword and cannon. Whilst Churchman, Papist, Presbyterian, Independent and Anabaptist, alternately fought and prayed, rose up and trampled each other down; whilst Cavalier and Roundhead, Charles the First and Prince Rupert, Fairfax and Cromwell were crimsoning with fraternal blood the fields of Marston Moor, Naseby, Preston, Worcester and Dunbar, to establish a kingdom or a republic that might last but for a day, these obscure people were inaugurating a real, though silent revolution which (only partially consummated it is true) has yet overturned thrones, freed millions of slaves, and in the practice of national arbitrament may yet put an end to all wars and fightings,

"'Till the war-drum throb no longer, and the battle flags are furled
In the Parliament of Man—the Federation of the World!"

It was time for a new dispensation or a revival of the old in its purity. Out of all the ecclesiastical rubbish of the age, the débris of former systems and superstitions, what was there left to build on as from the beginning? Simply the reason—the soul of man—and its author. Tradition may fade away in the lapse of time, the written word may be misinterpreted or falsified, but God will not leave us without the living witness in our hearts. If we will be watchful and obedient to this light of truth, which is the Divine, the universal reason operating on our hearts and consciences, it will in time lead us into all truth; our spiritual perceptions growing day by day as we conform our lives to their teachings. The spirit of God that inspired in the past is still universally present in the hearts of men, and Revelation is not a closed book. Such was the faith of the Quakers!

Turn we now to the founders of this faith. We might classify the three who stand out prominently as the great leaders in its origin in English history, the three great lights (besides the inner light) of Quakerism: George Fox, the prophet and preacher; Robert Barclay, the scholar and writer; William Penn, the colonizer and statesman. Not that this classification is complete and distinctive, for there was a blending of different characteristics in all. Though pre-eminently a teacher of religion, Fox was also a vigorous and prolific writer and an executive framer of no mean order, for chiefly he originated the policy and government of the society. Robert Barclay, of Uri, the descendant of a noble, old family, memorable in Scottish history, was a noted preacher and propagandist, though decidedly the scholar and most able literary defender of their principles; while the acute intellect and high culture of William Penn rendered him eminent in every field of labor into which his ardent enthusiasm carried him, though now chiefly known to the world as the successful founder of the great Quaker Commonwealth of Pennsylvania. Of these three, time allows but a brief glance at one, and it will properly be directed to the great and original genius, George Fox, the founder. Born in 1624 in Leicestershire, the son of an honest weaver called by his neighbors, after the fashion of those days, "Righteous Christer," and his mother of the stock of the martyrs, he was from early childhood a grave, serious lad; faithful, earnest, conscientious, pure and delicate minded, shunning all evil associations. His friends thought the career of a minister indicated for him, but were persuaded otherwise, and he was apprenticed to a shoemaker, who also fed and dealt largely in sheep. His employer entrusted much business to him, which was faithfully transacted, and it was well known that when George Fox said "verily" there was no moving him. At the age of nineteen, noting the inconsistent conduct of life of professedly religious people, he fell into deep trouble of mind over the vanities of the world, and having cried unto the Lord, he was commanded to forsake and become a stranger to them all. Obeying the voice, he left his relatives and friends, and for years wandered to and fro, often in the fields, woods and solitary places of the country side, and sometimes buried himself in the greater wilderness and loneliness of London. He fasted much; he meditated and prayed; he longed for a knowledge of the great mystery of existence, its aims and ends. He sought knowledge of the clergy. One advised him to marry; another to join Cromwell's army; still another would have him be bled and take medicine. (In those days even the ministers would give physic to a mind diseased.) Failing in obtaining light from the churchmen, he turned to the Dissenters, who seemed earnest and zealous, but soon became convinced that they were themselves still in darkness and could not speak to his condition. Saddened, he sought solitude again, and struggled alone with the problems that were pressing on him for solution. Sometimes, breaking through the gloomy clouds that encompassed him, gleams of celestial light shone in, irradiated the dark places of his soul and possessed him with a heavenly joy. In such seasons of retirement from the active world, amid conflicting and alternating temptations and exaltations of the soul, the great religions of the world have been born. At last, to the earnest seeker, amid the calm and silence of nature, falls upon the waiting heart "the still small voice" of God.

So it seemed to George Fox that, "through many alternations of hopes and fears, his seeking mind was gently led along to principles of endless and eternal love." He was taught that "there was an authority in man to teach him that God would himself instruct his people without the intervention of university-taught priest; that none can be ministers of Christ but in the Eternal Spirit, which was before the Scriptures were given forth; that there is no holy ground in church or temple, but only in people's hearts." Confident of the truth of revelations that had given peace to his soul and gladness to his heart, he felt impelled to go forth and proclaim the glorious truth to the world. In his own language: "Thus travelled I on in the Lord's service as the Lord led me! I was to bring people up from all the world's religions, which were vain, that they might know the *true* religion—might visit the fatherless, the widow and the orphan, and keep themselves from the spots of the world. I was to bring them up from the world's fellowships, prayings and singings, which stood in form without power; from Jewish ceremonials; from heathenish fables; from men's inventions and windy doctrines, which blow people about from sect to sect; from all their images, crossings and sprinklings, with their holy days so-called, and all their vain traditions which they had got up since the Apostles' days."

In other words, George Fox was to preach a religion outside of churches, rites and creeds; a gospel of humanity, whose seed was universally implanted, and destined to grow and blossom and fruit into the practical, worthy living of lives patterned after that which Christ led upon the earth. That God is the universal loving Father and all mankind one brotherhood, was not merely a fine theory with George Fox. In every act of his life he sought to give that practical exemplification of his belief, which is usually the most obnoxious form in which we can express our radical ideas.

Sir Henry Vane was an advanced republican of those days, but he could treat Fox with severity for not removing his hat in "his honor's" presence. On the other hand, it reads rather singularly, in the wonderful Journal, when we often find the great historic "Protector" of England referred to as simple "O. Cromwell." The title of "Friend" was the address adopted universally by the Quakers to emphasize the great principle, "all men are equal by their birth." Whilst reverencing the Scriptures, Fox freely denounced the Bible idolatry so prevalent in his day, and not totally extinct in ours. As he heard the bells of Nottingham Church, near the home of his boyhood, calling the people together, "the sound struck to his heart, for it was like a market bell assembling them for the priest to offer his wares—the Scriptures—for sale." One First-day morning he was moved to go to the great "steeple-house" and cry against their idol. "When I came there," says Fox, "all the people looked like fallow ground, and the priest, like a huge lump of earth, stood in his pulpit above. He took for his text those words of Peter, 'We also have a more sure word of prophecy,' and told them this was the Scriptures, by which we were to try all doctrines, religions and opinions. Now the Lord's power was so mighty upon me and so strong in me that I could not hold, but was made to cry out, 'Oh no, it is not the Scriptures; it is the Spirit.'" This was hte seed of a great religious revolution. As a great writer has expressed it: "The Bible enfranchises only those to whom it is sent; Christianity those only to whom it is made known; the creed of a sect those only within its narrow pale. But George Fox, resting his system on the Inner Light, redeems the race." On the hills of Yorkshire he had a vision of the great work of God in the earth, seeing the people thick as motes in the sunbeam that should be brought home to the Lord, that there might be but one Shepherd and one sheep-fold in all the earth. Possessed with such enthusiasm, no wonder that neither raging priest nor stoning populace could daunt him! On he went in his divinely-appointed mission, and as he rode "the seed of the Lord sparkled about him like innumerable sparks of fire." The clergy, who at first provoked discussion with him, soon learned to shun the unequal encounter; they trembled and went away at his approach, and "it was a dreadful thing unto them when they were told that the man with leather breeches was come." Through fiery trials, the sect that he founded (I term it a sect, though a creedless one) grew and prospered and multiplied into thousands and tens of thousands. In fact as many as 4,000 of them were at one time in prison. At last, a part found a home in the New World, and fair Pennsylvania offered a peaceful refuge not only to themselves, but under their laws to the persecuted of every sect, where every man was secured the right to worship God according to the dictates of his own conscience.

THE QUAKER WEDDING.

No wedding bells rang out in air,
    No strains of music blended,
No orange blossoms decked her hair,
    Nor bridal veil descended:

The bridegroom to the bride gave naught
    Of symbol ring or token,
No rite was with tradition fraught,
    No churchly vows were spoken.

With all that ritual imparts,
    Performed with pomp and splendor,
No tie unites two loving hearts
    Like simple words and tender.

It needs not old cathedral rare,
    Nor tones of organ pealing,
To sanctify the pledge they share,
    Or wake the soul's deep feeling.

So "in the presence of the Lord"
    And loving friends around them,
Their own lips spoke each solemn word
    That, life to life, hath bound them!

TWO SCHOOLS OF FICTION.

AS THE era of Phidias was the golden age of Sculpture; as the period of the Renaissance comprehended the palmy days of Painting; as the age of Elizabeth signalizes ever the glory of the Drama; as the first half of the nineteenth century stands eminently prolific in Poetry; so the present, its latter half, is distinctively the era of the Novel.

If the harvests of the Imagination in Literature have heretofore been garnered from the fields of Song, either we are now suffering the reaction and rest of nature in a period of dearth and famine, or else, while those fields of Poesy lie fallow for a season, the reapers and the wagons have been diverted to other fertile plains—the fair, broad acres of Romance.

Is it a question of alternate seven years of famine following closely the seven years of plenty, or, more haply, a matter of rotation of crops in literary agriculture? This is a question I still leave with the reader for final determination, simply expressing, in passing, my humble opinion that Imagination and Fancy are yet neither dead nor sleeping, but daily walking abroad among the children of men.

* * * * * * * * *

Nothing is more wonderful than the rapid development in this age of the idea of the Novel—its scope and its responsibilities. Formerly its function was but to amuse and entertain. The tales of fiction were like those of the Arabian Nights or the Decameron—recounted by some smiling, light-hearted, light-tongued Scheherezade or Boccacio—the careless "singer of an empty day."

Now its province has been immensely extended. One outlying principality after another has been annexed, until it has come to embrace pretty much all the literary kingdom, and with this idea of its increased scope and power, comes that also of its added duties and responsibilities. Good fiction "is profitable for instruction, for reproof, for doctrine," and, like the "perfect woman" of Wordsworth, it should be

"Nobly planned
To warn, to comfort and command."

With all the burning questions of Biology, Psychology and Theology pressing close upon us now in our daily lives, we need all the added illumination that the sidelights of Literature can throw upon them; hence we are inclined to welcome the tendency of jurists and divines to gather texts for decisions and sermons from the master-pages of fiction.

The Study of Fiction has, indeed, become of such prime importance that, together with the study of "Shakespeare and the Musical-Glasses," Fluxions, the Binomial Theorem, the "Fourth Dimension" and Italian Renaissance, it should be included in the Curriculum of all our Public Schools. In making this suggestion, I by no means blink the recognized fact that the schedule of studies pursued by our youth is already so extensive as seriously to overcrowd their time and capacities. In order, therefore, to make room for these modern and more important studies, I would lop off several of the old-fashioned and nearly obsolete ones, beginning with "the three R's!"

It was said by Shakespeare—or the other fellow—that "all mankind loves a lover." And somebody else has remarked that no novel or romance is complete that does not embrace the subject of love. We are then enabled to assert that all mankind loves a good story. It is pretty safe to assume this. The taste for a good story, together with that for a good dinner, is, perhaps, more nearly universal than any other. ."We may live without poetry, music or art"—in fact, a great many people do get along without any of these very passably indeed—but this age is pre-eminently one of story-writing and of story-reading enjoyment. Some one has asserted that even judges of the highest courts recreate themselves by reading romances, and a judge of the United States District Court in Kansas recently showed his familiarity with one romance at least by citing from the strange case of—"Dr. Jekyll and Mr. Hyde." The taste for and the habit of story-reading is spreading to be almost universal. The leaves of the paper novel are shed abroad far thicker than the "leaves of Vallambrosa;" they fall alike "on the just and the unjust." The cabinet officer, and the elevator boy who may one day rise to the top story (one story beyond the high official) and come to own a "cabinet" of his own—both of these alike indulge in the relaxations of romance. The library of every household swarms with the *Household Library*. When we go out we find that the *Franklin Square* lays over every little park in city or town, whilst *The Seaside* pursues us through every summer resort, and follows us into the fastnesses of the mountains.

And conversely, while directly in obedience to the law of demand and supply, about everybody who does not read novels (except, perhaps, the critic) has gone to writing them. No literary reputation is complete now-a-days unless it has been topped out with a romance. All the poets and poetesses finally abandon their verse, proceed to join the less tuneful choir, and swell a chorus not their own.

Possibly romance pays the best. Undoubtedly it pays the best. "A rose by any other name would smell as sweet," but had the two Roes-es—he of the Hudson, and he, the smartly imitative Chicago namesake—got themselves planted in the gardens of poesy, instead of realistic romance, they might long ago have wilted on their stems, lacking all sweet savor of interest to reader or publisher. All the authors are turning to the novel. Even Amelie Rives Chanler has tried her hand at it. Senator Ingalls has been about it for a long time, but unfortunately—or fortunately—for the world and his fame, the manuscript, like the first of Carlyle's "French Revolution," went up in the flames. The luckless man who writes—or buys—a poem now-a-days may come to find himself in the category of the doubting lady I met in a picture store the other day. She had had thoughts of buying an etching, but she was afraid they would "go out!"

The novel is fast assuming, like Lord Bacon, to take all knowledge for its province. If a man wishes to propound a new fact or a new theory, if he has explored a new continent or a new province of thought, he hastens to exhibit it to the world, thinly disguised, it may be, under the veil of fiction.

If his style be passably attractive, he wins a hearing for himself and his pet, and an audience of readers probably ten times as great as if he had addressed himself in the old-fashioned method, to his special clientèle.

The youth, at least, of this country, have hitherto

learned far more about the interior of the "Dark Continent" from the pages of Thomas W. Knox, (in "The Boy Travellers,") than from the books of Livingstone and Stanley, while, of maturer minds, hundreds to one prefer to read "Robert Elsmere and "John Ward," to tracts on "Miracle," or "Future Punishment." Charles Dickens assailed Yorkshire schools, imprisonment for debt, Chancery courts and Gradgrind Realism—while Charles Reade attacked the abuses of insane asylums, trades-unions and the prison "solitary system," from the vantage-ground of fiction, with correspondingly wonderful effect.

Who would not choose to write the romantic story of "Uncle Tom's Cabin," and thereby arouse a whole nation to feeling and action, rather than to deliver tracts and facts to people of previous anti-slavery antecedents?

* * * * * * * * *

A few years ago, William Dean Howells, himself our leading story writer, took it upon him to issue, as one having authority, a proclamation to all the inhabitants of the land of Fiction, or those who should come to sojourn therein. It comprehended a notable announcement, and was, at the same time, both a requiem and a pæan of rejoicing. "Great Pan was dead"—the days of Romantic Fiction were over! For many days it had, indeed, been nothing but a farrago of romantic bosh and hash; now, peace be to its hashes!

The past of fiction should be as "a tale that is told;" in fact its stories—the stories—had all been told. There were left no new plots and incidents—all had been exhausted. Now they were simply being rehashed over and over again by the story writers. Invention had reached its limit. Scott and Dickens and Thackeray, and all their followers were obsolete or obsolescent. It was time for a new dispensation; there *was* a New Dispensation, that of wholesome Realism, and Henry James was its prophet—of whom, he, Howells, was proud to enrol himself as the first disciple!

This pronunciamento was duly followed, from time to time, by others from his pen, elaborating and fortifying his theory, and ably expounding the gospel of Realism. Also he began to cast his nets around in the deep and shallow waters of Literature, and successfully and successively to land several realistic fish of greater or less degree, some of whom proved literary leviathans in breadth, and others "very like a whale, indeed:"—Turgénieff and Tolstoi, of Russia; Flaubert, of Paris; Valdez, of Spain; and E. W. Howe, of Atchison, Kansas.

And the literary world was, for a time, very willing to give some credence to this new creed, and a patient hearing to its advocates. In fact, we had previously grown a little tired of the romantic school, since most of its great masters had passed away. Its second rate fellows scarce charmed us at all. They possessed not "the grand style." What in their works was old, became too trite in their hands. They had exhausted most of the old-time plots and devices, and especially all the old life-saving apparatus, by means of which the hero rescues the heroine and gains her eternal gratitude and love. All such had become too frightfully familiar and worn to be serviceable any longer. On the other hand, their new tricks and spasms, their violent exaggerations, distortions and monstrosities were apt to be so wild and fantastic as to strain entirely too hard upon our imaginations. These writers had developed and educated a class of novel readers which is one of the special characteristics of the age, the habitual novel-reader. A class which loves stories for their sensationalism, and lives upon it, as an opium or hashish-eater lives upon his drug and its excitement, having lost relish for anything more wholesome and nutritious.

This class, which still abounds in every community where there is a circulating library or a "news-stand," is constrained by the force of habit and the condition of mind induced by such reading, to rush pell-mell through one story after another, never halting to enjoy a charm of style or happy touch of characterization, but always in pursuit of incident, plot and sensation. Their minds, in time, acquire a sieve-like property that lets every story through, in turn, as fast as poured in.

But even the "circulating-library-fiend" was inclined to welcome something new. In fact, like the Athenians of old, they are always looking for some new thing. For awhile, everybody took their regular instalments of Howells, James, *et id omne genus*, the New School of Fiction, with zeal and apparent zest. Under the stimulus of the New Idea, wonderful results were reached, especially in the literary development of its two great exemplars. The world had long admired the early work of these brilliant writers; the purity of style, the wealth of observation, the keenness of analysis. In Howells, especially, one experienced a charm of freshness and originality, a delicacy and nicety of touch, united with a geniality of feeling, which was extremely fascinating. Beginning with his earlier sketches, particularly "Their Wedding Journey," and its continuation, "A Chance Acquaintance," this culminated in the "Lady of the Aroostook," a story which, simply and delicately told, without making any pretence of breadth of treatment, and exhibiting but few characters, yet discriminated them so finely and handled them all so admirably as to give fair promise of the highest literary eminence for the writer, in the near future. Has that promise been fulfilled? Let the descending curve that has led so swiftly adown the slope to the dreary flats of "Annie Kilburn" afford the answer! For Henry James, his "Bostonians" settled the question some years ago—and his partial struggles into pseudo-sensationalism of late will scarce retrieve him. If we should venture to render the historical verdict, it would be "Killed by a Theory," the theory that Realism in Literature is the only supreme good. Too much exclusive holding up a looking-glass before a very indifferent and common-place Nature!

—And the Romantic Fiction that should have been entombed by these men was soon on its feet again, livelier than ever!

A new crop of romantic story-tellers has sprung up like Jonah's gourd, and gained already a multitude of readers and lovers—"Called Back" Conway, "Strange Case" Stevenson, and that haggardest of blood-thirsty, story-telling fiends, Rider Haggard. Their recent popularity sprang into such immense proportions, as only to be accounted for, possibly, as a reactionary protest against the extreme and exclusive realism which had prevailed before.

But this later school of romantics is itself in turn too violent in force and too extreme in direction, and as such cannot long endure. In fact, just at present, it might seem as if we had entered a notable period of psycho-theologic fiction.

* * * * * * * * *

Since the above was written, Mr. Howells has published

a new story—"A Hazard of New Fortunes"—which, virile and inclusive in grasp and treatment, fulfills the promise of his earlier days, and places him conspicuously in the front rank of American writers. This exemplifies that in literature, as in metereology, an ascending curve may closely succeed one of descent, and regain all that had been lost thereby.

In view of his increased breadth and vigor, we shall scarce complain, indeed, that in this most "modern instance," he has left the realistic restrictions of inherent probability so far behind him as to bring together unappointed, from remote parts of the city to a certain street-corner in upper New York, no less than three of his principal characters, at the precise moment which awaited a "striking" denouement! Nor even that in another recent story, he should employ a literal *tour de force* of the most violent character, to relieve himself at once of his hero and a dilemma!

We cordially accept William Dean Howells as our Dean of Fiction in the University of Literature—but what especially pleases us in him, of late, is that he is so delightfully romantic. Not alone in his sentiment—he was often deliciously romantic in that before—but lately, in his incident as well!

* * * * * * * * *

A cursory examination of a few of the noted masters of fiction in the past generation may not be out of place, but must be very briefly taken. What part, conscious or unconscious, did this question of realism or romance have in the production of fiction in that era?

Evidently the drift of Sir Walter Scott's genius in fiction was essentially romantic. While he naturally sought epochs in history wherein the spirit of the time was congenially of that order—such as the era of the Crusades—or, at least, so far remote in the past as to allow full license to the imagination in treating them as such—as for instance, the Feudal ages—yet even when he lays his scenes in periods as nearly modern as those included in the last century, the Great Wizard is able to cast such a glamor of romance over prosaic times and humble events as to make them fairly glow in the magic of fancy—"the light that never was on sea or land." And yet, Scott has so deftly interwoven the warp and woof of fact and fiction, matched with the bright colors he could so skilfully impart; has gathered from old chronicle and legend so much that, whether true or not, bears the verisimilitude of truth, that his romances seem all infused with the hues of history, and if not "the very age and body of the time, its form and pressure," to be, at least, something so much finer, that we are more than willing to accept them in preference. While he, no doubt, illumines with a factitious glow some of the dark pages of feudalism, yet many of his incidents of story are no more essentially romantic, after all, than those chronicled by Princess Anna Commena or Philippe de Comines.

But it is not alone when "the pulse of life," in his magic creations, is "beating to heroic measure" that Walter Scott is truly great. When he portrays the fortunes and vicissitudes of that humble family of St. Leonard's Crags, the many homely touches of realism naturally incident to the daily life of Jeanie Deans, the cow-feeder's daughter, serve as a magnificent foil and charming back-ground to the noble purpose that animates her, and bring out all the more vividly, the glory of her achievement.

"The Heart of Mid-Lothian" is a conspicuous example

of what I consider the higher order of fiction—that in which fact and fancy, realism and romanticism mingle and harmoniously blend. If I were asked to cite a modern example of this felicitous combination, perchance the first one that would rise to mind would be "Lorna Doone," the single notable story that has yet been written by R. D. Blackmore.

It is to this commingling of the two—though so often impaired in happy effect, by trick and exaggeration, by crudities and quiddities of characterization and vicious mannerisms of style—that may be attributed, as I deem, the popularity of Charles Dickens as a novelist. It would be difficult to tell whether he is most Idealist or Realist. No writer was ever more of the latter in his close observation and reproduction of unusual and odd types of character, but in sentiment and direction he was always thoroughly romantic. Had I time, I would gladly go into this further, by way of illustration, but the numberless examples which might be adduced in proof will readily occur to all familiar with his writings.

Thackeray was another great master; great in style and great in matter, because equally at home in either school of fiction—the two not exemplified so often by being intimately blended in the same novel, as in the pages of his great cotemporary. To give typical instances, I would cite "Vanity Fair" as one of the most characteristic examples of a realistic novel ever written, with "Henry Esmond" and "The Newcomes," leaning closely to the romantic school. In all fiction where is there portrayed a finer character than Colonel Thomas Newcome—so ideally grand and yet so essentially human in his imperfections!

In the same connection, let me revert again to Dickens,

to cite the "Tale of Two Cities," as evincing most assured power as well as strength of style of any of his romantic tendencies, with "David Copperfield," as, perhaps, his finest union of the two schools—the realistic and romantic.

Possibly the greatest name in purely romantic fiction, of the last generation, is that of Victor Hugo.

In the old-time deadly conflict, between classicism and romanticism in the French drama, his was the embodiment of the latter idea, and his the early triumph as its representative champion. Thenceforward, the contemner of the "Third Napoleon" was consistently of this school of literature. In Poetry, in Fiction, and in Fact (that is, as nearly as the great Frenchman ever touched an actual, prosaic fact) Victor Hugo was a Romanticist. In spite of its crying—and shrieking—defects of style, bordering often close upon the hysterical, "Les Miserables" remains one of the most wonderful products of imaginative genius.

Charles Reade probably considered himself a realist of the first water, and indeed his style was eminently such, being arid in the extreme. His ideal of construction was also the acme of realism; it being his habit to accumulate first a mass of raw material, clippings from newspapers, and other sources, of such reported facts as struck him as available for novelistic purposes, especially if they bore on some pet theory which he desired to exemplify. These he would arrange in scrap-books on some system of Index Rerum. He had a veritable "Gradgrind" penchant for facts, and he had likewise a genius for utilizing them as a statistician does figures, so that they would tell a marvellous story—any kind of a story he wished. When this process of coloring and distorting the facts begins, he changes, chameleon-like, into the Romanticist, pure and simple. His characters, too, always act on the romantic plan, instead of that of natural development. His women, especially, can always be relied upon to do just the thing they logically should not; it being one of Reade's pet theories that all women are prone to act "like Paddy's pig," and that, in consequence, whenever you want them to go to Bantry you must pretend they are on the way to Cork.

Both Reade and Wilkie Collins, however, are sufficient refutations of Howells' theory, that "the stories have all been told;" unless, indeed, he should put in the saving plea that such were among the rush of creditors who drew out the last coin, and left the bank of invention bursted.

Both of these kept up telling interesting stories to the last, and Collins had always a big fund of incident, though his style, like that of the dictionary, may seem at times "a trifle jerky."

The mention of Charles Reade's scrap-books reminds one of Hawthorne's "Note-books," published after his death, and very unwisely, perhaps, for his best fame. We should have preferred to imagine those wonderful romances as projected, lava-like, from the surging and overflowing crater of a passionate imagination fused and glowing at white heat, for they seem veritable products of secret seething recesses of heart and mind. Instead of this, the note-books take us into the romancer's larder and kitchen, where the apples, raisins and meat are being accumulated and chopped, and where the suet is being tried out and the brandy decanted; all of which, skilfully compounded and cooked up, should duly eventuate in the mince pie of romance, that whilom we had found so toothsome.

In his elaboration of detail—often wonderfully minute, sometimes apparently trivial, but all tending to heightening of artistic effect—Hawthorne is a Realist. In his choice of out-of-the-way characters and unusual aspects of life, and their development into startling and tragic events, he is a Romanticist—but chiefly, in his study of the mysterious workings of mind, especially of the morbid and abnormal sort, he is a Psychologist. While a predominance in importance may be granted to the latter function, it is probably owing to the masterly blending of all these characteristics that we cede him high rank in literature.

Eminent in differing departments of fiction, striking its leading chords of realism, romanticism or idealism to produce some of their finest tones at will, the psychologist in George Eliot broadened out at last into the philosopher, whose subtle and profound reflections make us sometimes oblivious of the fact that it is a novel, a work of the imagination, that we are perusing. And yet, as the gifted writer carries us along with her in these excursions into the realms of deep thought, we seem to leave behind that ideal country wherein we first journeyed, and our guide loses somewhat of the artist in taking on so much of the metaphysician. For this reason, the earlier "Adam Bede" and "Felix Holt," possessing more romantic and human interest and using more direct story-telling power, strike me as altogether artistically finer than her later works.

To leave the high table-lands which the mind of George Eliot inhabited, and come down to the actual misty headlands of the Atlantic and the wild shores of the Hebrides, William Black is one of the most agreeable and popular novelists of to-day of the romantic school;

writing always with graceful pen "dipped in the hues of color" and of fancy—

"Tingeing the sober twilight of the present
With coloring of romance."

But the atmosphere which should reflect the glow and the color is apt sometimes to become rather attenuated, being palpably evoked more through force of determination of the writer than the power of assured imaginative genius.

On the other hand, imagination itself, though a kingly power and prerogative in a romantic era, may scarce be safely substituted in picturing those types of a modern age wherein the correct results of the faithful observation of realism are imperatively required. As instances of imaginative pseudo-realism, one might cite Cooper's Indians and his Leather-stocking hero; Bret Harte's California miners; Cable's Acadians and Creoles, and William Hardy's Shakesperian rustics of to-day. These are all interesting as mental projections of genius—but they fail as veritable human beings.

* * * * * * * *

Do novels of the Romantic school, as claimed by Mr. Howells, uniformly give us false views of life, influencing us to illogical, impractical and harmful course of action? If so, his indictment of the school has great force. Possibly this objection may properly lie to much of the sensational literature of the day. It is hurtful to the minds alike of youth and children of a larger growth. But the assertion that all romantic literature is injurious is as untenable as might be the converse proposition that all realistic writing is beneficial—or that a chess-player's mind is so strengthened and disciplined by his pursuit that he may never make a false move in the game of life!

If nothing but realism be proper food for mind or soul, if imagination and fancy are to be wholly ruled out, let us drop professed fiction and rely solely on the daily newspaper for our mental nourishment. That is realistic enough in all conscience, and imaginative enough also, perhaps, during political campaigns. I would, however, maintain that romantic fiction of a high order may be most serviceable to us—even toward our apprehending life in many of its important relations.

Oculists tell us, and experience confirms it, that it is better for us to have two eyes, and use them—even though one or both may be imperfect—than to rely on one alone, however complete its visual capacity. Somehow in the correlation of use, the faultiness of either will be measurably rectified, and we will have truer apprehension of the proper relations of visible objects.

In like manner, it will be a gain to us to look at life through the window that the novelist opens to us, as well as through that of our every-day experience. It is well for us to contemplate the differing aspects of our existence; the usual and the unusual; the every-day and the gala-day; the near and the far. So shall we gain a truer vision, a broader outlook, a fuller comprehension of that complex landscape of the soul which we call human life.

William C. Gannett, in that most helpful of all "practical" sermons, "Blessed be Drudgery," has enjoined upon us this lesson, as the summing of the whole philosophy of contentment: "If you cannot realize your ideals, then idealize your reals!" This is what the best novels—the combination of the romantic and realistic—may help us to do: to "idealize our reals."

* * * * * * * *

In one of his earlier poems, Bret Harte pictures to us

a California mining camp on the slopes of the Sierras, in the early days of the excitement and thirst for gold. Weary, worn and haggard from the toils and anxieties of the fierce pursuit, one night, around their camp fire, they listened to a younger member of the party, who produced from "his pack's scant treasure," a "hoarded volume" of Charles Dickens—the Story of Little Nell:

> "Perhaps 'twas boyish fancy—for the reader
> Was youngest of them all—
> But as he read, from clustering pine and cedar
> A silence seemed to fall;
>
> The fir-trees, gathering closer in the shadows,
> Listened in every spray,—
> While the whole camp, with Nell—on English meadows,
> Wandered and lost their way.
>
> And so—in mountain solitudes,—o'ertaken,
> As by some spell divine,
> Their cares dropped from them—like the needles shaken,
> From out the gusty pine."

Such is the spell of the *Imagination*, and such the compensations and solaces it sometimes bestows on the rudest and hardest of lives. What realist will dare deprive us o them all!

TWO TRAVELERS OF SUCH A CAST.

THE Psychologist and the Realist of literature sometimes walk its great highways, side by side. They prefer to travel its dusty high road, because that is lined with the abodes of human life and crowded by its great throngs— and it is life, with its experiences, that is the object of their pursuit. It is that alone which is worth their study and admiration. For them, there is neither pleasure nor profit, neither poetry nor worthy prose in nature, unconnected directly with some individual, or type of the species Man. For them there is no "pleasure in the pathless woods;" there is no "rapture on the lonely shore;" there can be no "society where none intrudes, by the deep sea"—nor "music in its roar." In fact, it may well be doubted whether, in absence of human ears, there can exist any "roar" at all!

What a blunder in fact, and how utterly unpoetic in idea, those lines of Gray:

"Full many a gem of purest ray serene,
The dark unfathomed caves of ocean bear."

How could they bear them, even in poetic fancy, so long as the caves are "unfathomed" by man? To these two confrères, there is no landscape or marine, however full of beauty or of grandeur; of mystery, of sentiment or of subtle suggestiveness; no canvas of Claude, of Rousseau or Corot, that the foolish world of fancy had been wont to term "poetic"—unless there haply be a piece of "genre" painted as planted in the fore-ground.

Never in sublimest poetic fancy could they hear the "morning stars sing together for joy"—unless they, or some prototype of theirs, should be on hand, with sharp ears to hear, and a sharp pencil to take down the notes!

With one common aim, expressed by the phrase "to hold the mirror up to nature"—that Nature comprehended by the unit Man—these two travel together, along the highway for awhile, observing and noting the types as they pass them by. But the pace of the Realist is really the swifter, inasmuch as he is satisfied by catching the reflections of the looking-glass that he holds up, or the camera that he trains upon his victims. If he catch a fair and life-like image, which shall give representation of outward appearance, with all the characteristic attitude and gesture and individual action, the Realist is well content. Not so, for any length of time, the Psychologist. His work goes far deeper and requires more time. He aims at complete analysis of the inner as well as outer self. He fain would dissect the "subject" completely, lay bare every quivering nerve, and turn the compound-microscope of critical observation upon each corpuscle of blood and every fiber of the heart, detecting the nature and course of the life-current, and every hidden spring of action. This takes more time than realism has to give, but who can doubt that the results are more complete! When the Master is through, the process has been most thorough, indeed! You can turn the dry-bones—all that is left—over to the Scientist now, if you can possibly find one so wholly devoid of imagination that he cares for nothing but to articulate skeletons. But be chary of terming this process of the anatomization of humanity Poetry—or conceiving that there is much poetry in it, much less of romance! When you have got at the heart of all mystery in man's soul—and plucked all the mystery out of it—the poetry vanished just before; for without mystery, imaginative genius has no atmosphere to work in, and perishes like the animal in an exhausted receiver.

* * * * * * * *

There are some almost intangible essences constituting the boquet of a fine old wine, that may apparently be analyzed, but not synthetized. Possibly there is some subtle and evanescent element, more ethereal than the ethers themselves, that exhales and escapes in the process of analysis; for when the chemist of liquor manufacture comes to combine his known alcohol and water, his sugar and gum, his acids and salts, his œnanthic, acetic and other ethers, in due proportion—the resultant liquor is something flat, insipid and wholly distasteful to the refined palate.

So with poetry; it is almost impossible to define it, for its very essence is liable to escape in the attempt to describe and circumscribe it. Rather than any substantive thing, it is a pervasive spirit that can be apprehended by our finer sense but scarce bottled up in any dictionary, or parcelled out in definite proportions by any exclusive critic. The poetry of synthesis—of rule and measure—somehow always fails, at the last, to include some of those vital elements which enter into its most ethereal composition.

The connoisseur in wines will readily detect the spurious wine of combination, and pronounce it "dish-water." He will discriminate nicely between this brand and that, and all the different vintages of each; he will have his own especial choice in brand and vintage, as each connoisseur may; he will invariably reject wholly the spurious article—but he will be very chary of claiming that *wine* is the product of any especial vineyard or soil, or clime.

And so, those of us who are neither critics nor connoisseurs, may still plead to have our liberty of choice left to us--fallible and faulty though our taste may be. What, ho! ye sound authorities in literature, "because ye are virtuous shall *we* have no more cakes and ale"— the toothsome cakes of Romantic Fiction—and the inspiring ale of Objective Poetry! "Yes," retorts our prohibitive Psychologist, "provided you take off the froth and take out the alcohol from your ale!" Vainly we protest that it is just the sparkle and the body to the ale that refreshes us. "No!" cries the Tetotal Realist, "cakes are fantastic food; there's no substance or nourishment in them. Meat is the only proper food for man!" And both shout in chorus: "Man is the only fact or factor in creation worth considering—and with water to drink and meat to eat, man is all right—and with a looking-glass to hold up to man and reflect his very form and pressure, literature is all right!"

Now far be it from this writer to controvert such theory, much less propound any theory of his own, save this one, simple and hard I think to refute, that in Literature anything is good that is first-rate! For the novel—the question is not, after all, to what school does it belong? Possibly the more schools it embraces the better, if all are properly composed and artistically disposed; if the picture have faithful realism in the near foreground, pleasing romance in the middle distance, and charming touches of idealism in the vanishing perspective—with the magic glow of genius irradiating the whole landscape! Provided the work is noble in design and artistic in execution, the main question follows: "Is it helpful?" Will it broaden the range of our sympathies? Will it give us a truer outlook! Will it teach us somewhat more of the nature of human relations and actions—the mysteries of human life?

* * * * * * * * *

But after all, for the mass of readers and with the mass of novels, their chief function must be to entertain, to divert, to soothe. How often the otherwise weary hours of pain are shortened, the pain itself somewhat assuaged, or at least for the time forgotten, by the perusal of a good work of fiction! How the mind, that had been drawn to its utmost tension, has been healthily relaxed through the kindly influence of the fascinating page! By the way, what a compliment we think to pay to some work of history, of travel, of science even, happily treated, when we say, "it is as fascinating as a romance!"

For all of us fast slipping over that imperceptible boundary between middle-life and old age, whether care-worn men of affairs, or weary and exhausted students—with nerves not quite so equable, nor spirits so buoyant as of old—when we come home at evening, fretted, troubled, well-nigh overburdened with the worrying realisms of life, and needing relaxation and enjoyment, we go into our library for a book that shall distract our minds and lead us into another and pleasanter realm of thought: we will take down—not "one of the grand old masters," not any of "the bards sublime" but—a good novel! Then—

> "The night shall be full of music,
> And the cares that infest the day
> Shall fold up their tents like the Arabs,
> And silently steal away."

THE REALIST IN ART.

WERE we to credit the current claims put forth, we might suppose the taste for music, and that for pictures to be well nigh universal. In Art, everybody professes to love the one and admire the other. Yet Shakespeare seems to have had some ideal monster in mental projection, who, untouched "by concord of sweet sounds," was only "fit for treasons, stratagems and spoils," and occasionally, some extremely candid individual confesses that to him there are no "odds in pictures." It may be admitted, I think, that there are different degrees of liking for art, with only a limited few aspirants whose liking is sufficiently a matter of true taste and ardent feeling to be justly ranked as loving.

The person who can contentedly pass through a gallery of masterpieces of art, giving just five minutes to the Sistine Madonna, three each to Correggio's Holy Night, Carlo Dolci's Saint Cecilia and Battoni's Magdalen, two to the Rembrandts, and an average of the same time to each of the twenty-one cabinet galleries; thus doing the whole first floor of the Dresden gallery in one hour; then hurry off to devote equal time and attention to the Grüne Gewœlbe—this traveller can fitly be handed over to the cruel tender mercies of the "personally conducted," without any fear that his ardent longings for art may suffer any conflict of torn emotions, "the pangs of despised love," or "the sickness of hope deferred."

The appreciation of such a man for art is of a very low grade. He views pictures chiefly from idle curiosity, or because he has heard it is considered the proper thing to do. With another class, the sentiment is far above this, and yet, it may be doubted whether their enjoyment of paintings is not the same as it so frequently proves in music, chiefly a matter of association. It is often the words of a song, the beautiful ideas they express or suggest, that appeal to us, rather than the music itself, though, through the union with such words, the music may come to be loved thereafter—calling up as it does by the subtle suggestion of association, the words that had moved us before;—the two having become intimately blended in our minds through that happy marriage of "perfect music wed to noble words."

Many old ballads possess this union in such degree that it is difficult to analyse closely and say wherein the chiefest charm consists. Mark the great singers on their *encores*—how naturally they strike the ballad when they wish to please the cultured and the uncultivated ear alike! Is it the music of them—or "that touch of nature which makes the whole world kin" in the words—which gives their grand success.

So in painting. How often it is the story told by the picture that really arrests and charms the average beholder, rather than the depth of tone and harmony of color in the picture itself! An old-time joker used to recoup himself whenever any of his funny stories "missed fire," by immediately certifying them to his obtuse auditory as absolute fact in every particular! In like manner you can often awaken interest in the mind of a realist when the picture fails to enlist his attention on the score of Art. If it is, for instance, an ideal girl's face—one of

Jacquet's, opulent in sensuous beauty, or Greuze's in "spirituelle" grace—tell him it is a portrait of Nilsson, or of a princess of the English royal family! If this does not answer, the only resource is to enlarge upon the gilding of the frame.

Engraving or etching may "tell the story" about as faithfully and successfully—and, in fact, these are enjoyed by many fully as well as the finest pieces of color. This especially applies also to a large class who have an almost purely *intellectual* appreciation of Art.

For the foregoing reason, a landscape which is a faithful transcript of an actual scene will be far more attractive to many than any ideal one, however full that might be of the higher charms of imagination and of feeling. A "sketch from nature" almost invariably "draws" with this class of minds if, indeed, it be not too plainly "out of drawing." However indifferent they may have been to it as a work of art, they become interested in it as an actual scene—particularly if it represents one they may have themselves beheld.

These people are Realists. They enjoy photographs of people and places—especially of people and places that they have seen or read about—and cannot comprehend why the photograph of a painting, skilfully colored up to imitate the original, is not as good as the original itself. By the constitution of their minds, there is no good reason, outside of its repute and religious association, why Murillo's "Immaculate Conception"—an ideal picture of the Virgin whom they have not seen—should interest them half so much as a good photograph of some loved relative or friend.

English artists, as a school, are famed for their devotion to the above idea. Their pictures must all "tell a story;" that is, represent something that either is, or purports to be, real in its essence—happening or that might happen—something in either actual life, history or literature. The artists of this nation paint for a practical people, and know what will sell in their market. We freely admit that their stories are usually pure, and often infused with a pleasant or touching sentiment. The Royal Academy catalogues show a large proportion of this kind of work, done very well, indeed. It is genre painting of a very refined character; comparing most favorably, in subject and sentiment, with that of the Dutch and Flemish schools of the seventeenth century, though far inferior in strength—as, indeed, it is in quality of drawing and brilliance of technique—to that of the French realistic school of to-day. In moral tone the English artist of this type is immeasurably superior to his Gallic neighbor. Lacking, however, the finer impulse and sway of the imagination, the realistic school of all nations fails to attain and inhabit the higher realms of art.

* * * * * * * * *

As a rule, we Americans, in common with the bulk of mankind, are realists and like realism in our pictures. If they fail to reproduce things just as they are in ordinary life, what are they good for? Pictures and statuary are representations—and if they do not represent things in their usual aspect and as they appear to our ordinary apprehension, what use have we for them? *What's the use?*

A poor dealer in merchandise in a certain Kansas town who had ventured to lay in for holiday trade, some store of bric-a-brac—what the Yankee travelled abroad called "articles of virtue and objects of bigotry"—was bored almost to verge of desperation by the invariable, persistent query propounded by every customer to whom he exhibited them: "What's the use of this?" "What's that for, anyway?" He found that he could scarce hope to sell them unless he suggested practical uses for them; and so, rather than have all left on his hands, he was driven to inventing various applications for them, undreamed of by their designers!

> He turned fine art into a sell,
> —And then he sold it very well."

These people were Realists of the first water. Emerson spoke to their apprehension an unmeaning parable in his

> "If eyes were made for seeing,
> Then beauty is its own excuse for being."

Keats wrote for them in a tongue as unknown as Sanscrit or Choctaw when he penned,

> "A thing of beauty is a joy forever,
> Its loveliness increases—it will never
> Pass into nothingness; but still will keep
> A bower quiet for us and a sleep
> Full of sweet dreams and health and quiet breathing."

This quotation, by the way, is from a book in my library, long ago appropriately bound in sheep by some Eastern realist publisher, who gathered three poets into one volume and lettered it

> "COLERIDGE, SHELLEY, ETC."

Just think of the author of Endymion, and the Eve of St. Agnes, being labeled as Etcetera!

I have alluded to the fact that a large class, embracing many cultured people, find fully as much appreciative enjoyment in engravings, etchings and sketches in black and white as they do in the finest examples of color. Without derogating in the least from the artistic merit of work within this range—a scope which embraces, indeed, some of the finest achievements in art—yet I would raise the query whether, in such cases, one element at least is not lacking in the mental constitution. In an æsthetic sense are they not to a greater or less degree color-blind? Whether this be so or no, it strikes me that, failing to appreciate the tones of color in a picture, they lose a great deal of that largess of sensuous joy which nature distributes often with such lavish hand. It is true, indeed, that the needle or the burin, "skilful and touched with passionate love of art," may intimate, to some extent, the gradations of those tones in their effects of light and shade; but surely those glowing or melting harmonies of color, which are the triumphs of a great painter, impart a beauty and a glory that no scheme of black and white combinations can ever successfully rival.

Speculating on this matter, I have sometimes fancied that some men's minds are so constituted that they may be said to be "drawn in black and white." These are often great thinkers, great reasoners; cool, dispassionate, clear-sighted, illumined with the clear, white light of truth. They may become eminent jurists, illustrious scientists, wonderful logicians and metaphysicians, grand philosophers even; but hardly great orators, great novelists, great poets or great divines! *These* should have the endowment of all the color that is in the universe. Like the first class, they may possess all that is best in Realism—but they should be Idealists beside.

To return to our consideration of the realists, whose love for art we have dared to call in question; the writer, having ventured so far, is half inclined to go farther and commit the unpardonable sin of doubting whether, as a rule, they are really seized and possessed of any great admiration for nature herself!

Let us suggest that the love of landscape in nature and on canvas is apt to be reciprocal; that if we possess the former we shall naturally be drawn to the latter, when any good examples are afforded us; while even from glowing effects in pictures we may be instructed to discern more in nature than we had dreamed of before. Browning has noted, in poetic phrase, that we see things when painted which we miss in reality, while Hamerton goes farther, and reminds us that we sometimes have livelier, warmer and kinder sympathies at the call of the imaginative artist than the real world usually awakens in us; the revelations of the sympathetic artist carrying us farther into the realm of the ideal than we could travel, unaided by his inspiration.

We may first learn to love resplendent sunset effects from the painted ideal, but once having acquired their appreciation, we shall discover ten times as many in the evening sky as he who cares naught for pictures and to whom they are as to one who, "having eyes sees not."

One summer evening, some years ago, a train load of excursionists—chiefly members of a Western Legislature, and their families—was approaching the city of Denver. When within some fifteen miles thereof, the engine collided with some cattle, and one poor cow became so badly tangled up with the wheels of the locomotive that the whole train was stopped. The delay was such that about all the passengers alighted. From this slope of the Plains, the great range of the Rockies was finely visible, and, at the moment, the view to the west chanced to be one of scenic effect rarely equalled in any land. The sun, curtained behind a bank of cloud lying just above the mountain line, sent forth shafts of light that, toned by the mists hanging between the ranges, suffused the peaks with the most delicate, the most ethereal and yet the most vivid roseate glow imaginable. Their forms shone forth sharply defined, covered with rich masses of transparent rose-color; while northward, Long's Peak mingled by imperceptible gradations with heavier cloud-banks behind, and the flanks of the whole range toned into darkest and intensest blue.

Idealized and glorified thus by those two great painters, Sunlight and Air, the enchanted spectator could scarce realize that this wondrous ethereal vision was, indeed, that Titanic, primeval mass of giant mountains, the "Backbone of the Continent." With every deformity hidden, every harsh, rugged outline softened into flowing lines of grace, they might well have passed for that beautiful range of Carrara which the traveller sees rise before him in the plains of Tuscany—but with the white marble tinted as the rose—or for that vision of the Delectable Mountains which we beheld in our childhood through the imagination of grand old John Bunyan. Nature, fortunately, is not always realistic, but images herself to us oftentimes, robed in illusive diaphanous veils of atmosphere, or tinted with the thousand harmonies of color.

—And all this shining glory of the western sky was beheld at the time by perhaps half a dozen of the train load! The eyes of the remainder might possibly have beheld something of it had they not been too busy regarding the mutilated dead cow, which had finally been dragged out upon the plain. In this contest, between the Real and the Ideal, the former commanded as usual, the sympathies of the great majority.

FROM REALISM TO IDEALISM.

However differing in our views as to its proper province to-day, we have little difficulty in agreeing that Art finds its origin in Realism.

It is claimed by certain of the Evolution school that, in retracing the growth of the religious idea in man, they find its root in the grossest fetichism. In like manner, it may be admitted that art finds its beginning in the crudest form of realism. If the passion for pictorial representation be not inherent in the race, it certainly begins very early in the history of man. We can hardly trace him so far back but that we find some manifestations of this faculty. As with the barbarous races of the present, so with prehistoric man; his traces and relics among the bone-caves show that, whether partly imaginative or wholly realistic, he essayed to carve on bones from which he may have stripped the raw flesh to appease his ravenous hunger, some semblance in outline of the savage beasts by which he was surrounded. If in no other way, he showed even then his superiority over the animals from which he had become differentiated, for man is the only one that ever makes pictures. In all the ages since, this faculty has been developing; crudest in the crudest and approximating the domain of art as he ascended in the scale of mental development.

Whatever practical purposes it may have subserved in the outset (including the art of picture-writing, for instance,) the period has often arrived in the progress of its development when, as so supremely in ancient Greece, its realistic phase became largely subsidiary; the actual took on the mystic tinge of the imaginative and the ideal. Striving to find in itself a medium of expression, art advanced upon its interpretation of what was finest and best in nature by seeking an ideal type that might symbolize, if not express, an imagined perfection, grander and more beautiful than mortal man himself could exhibit.

Comparing the art relics of Greece with those of other ancient nations, we can realize how far the idealistic conception of the Hellenes and their pursuit of art for art's sake had projected them beyond the era of crude products of realism in art that marked the highest stage of Egypt and Assyria. The conceptions of these peoples were largely realistic; even their imagination could soar no higher than the embodiment, in part at least, of the coarsely material forms around them; so for statues of their gods they constructed abominations in ugly combinations of beast and bird—winged bulls—their highest idealization of the material forces in nature.

To begin with analogies from kindred arts, we might set forth that there is far more in Oratory than the command of rhetoric, with all its manifold figures, its sounding periods, and tricks of emphasis and gesture. The soul of that true eloquence which moves and inspires men until they are swayed out of themselves, includes something beyond all these, which can rather be felt than adequately defined. There is something in Poetry beyond "the chime and flow of words which move in measured file and metrical array." Music is not solely "a succession of rhythmic vibrations and their pleasing effect upon the sonorous pulses of the ear." And so in the Fine

Arts. The highest art in Painting, in Sculpture, and in Architecture, embraces something far beyond mere representation, even of what is fine in nature.

And, indeed, it is well to remember that nature herself, as some writer has fairly discriminated, "is not all loveliness, all grandeur, all magnificence by any means, any more than she is all beneficence." We contrast her perfections with her imperfections. Only a tithe of the scenes she presents are worthy of reproduction. Many of her creations are crude and commonplace; some of her aspects are even repulsive, while others of her ruder features she herself, in happy moods, idealizes. Then comes, indeed, the proper moment for the artist to transfer them to canvas! And there are yet others which, possessing some grand capabilities of interpretation, need all the idealization which is in the soul of man to conceive, the product of rich suggestions garnered up it may be from Nature herself, in past ecstatic moments. So the great painter, in sketching the present landscape, transfers to its features something far finer than appears to common eyes, imparting a grace and beauty born of inspiration and of memory—thus adding to all that is worthiest in the actual scene the grand suggestion of all that might have been.

In like manner, the great artist in his creation of ideal characters imparts to the lineaments of the living model his conception of what is most lovely or tender, pathetic or strong—the expression of all the emotions or passions that "stir this mortal frame," as in turn he may wish to exhibit them. However closely approximating the painter's needs, the model is, after all, but a lay-figure, which the Master arrays with the vestures of his own royal imagination. Pictures that represent only the model fail entirely to excite our sympathy or admiration. They have no power to move us; "there is no soul in them," and "painted from model," is so palpable that it might as well be lettered across their face. The difference between painting of this class, and that wherein the true artist has succeeded in imprinting his highest ideals of men and women, "beaming with love, thrilling with tenderness, radiant with goodness, ardent with fidelity"—this difference is immense.

Hamerton, in his "Imagination in Landscape Painting," assigns a most important place in art to this faculty of the mind. Its exercise marks in great degree the vast superiority of Idealism over Realism. It is his theory, likewise, that in Painting as in Oratory, the chief element of the success of the master is his power to command our imaginative sympathy. This he claims as the real secret of influence, and he instances its power in painting, by the example of a picture by Normann, in the Salon of '85.

It was of the Sognefiord in Norway—a salt-water loch, enclosed by precipitous mountains of bare granite, whose oppressive grandeur shuts out forever the distance and half the sky. In this inhospitable scene, entirely bare of trees or verdure, are a few wooden houses that suggest life, and the pathetic interest of the work lies in the sympathy we immediately feel for the inhabitants. How can human beings exist, says our imagination, in such a desolate solitude? The colony, however, is not entirely isolated—the artist has linked them to the world without by showing a little steamer making its way into the calm deep water, with a line of foam at its bows. The charm of the picture is its suggestiveness—like that of Boughton's "Return of the Mayflower," or his "Two

Farewells." Such pictures not only attract our attention, but hold it. We return to them again and again, drawn by their power of moving our sympathetic imagination.

And yet some great reputations have been built up by a conscientious practice of Realism, and faithful reproduction of models in conjunction with accessories and properties that smack of the theater rather than the domain of Art. That of Alma Tadema is, perhaps, a conspicuous example. His rehabilitations of customs and costumes of classic Greece and Rome, recognized as faithful in an historic and archæologic sense, are, after all, dry and soullessly realistic in their representation; while the idyls of Sir Frederick Leighton, with ideal figures and scene whose exact type might not be found this side Arcady, are yet so instinct with true poetic feeling that they seem very near to the heart of that Nature which is of all times and seasons.

Compare, too, the picturesque and statuesquely posed and strongly painted manikins of Gerome, Meissonier and all their coldly brilliant school, with the assured power and dignity of the figures in Couture's "Romans of the Decadence," and Carl Müller's "Call of the Condemned;" with the whirlwind rush and strength of Detaille's and De Neuville's battle-pieces, of Schreyer's Arabs, and Schelmonski's or Kowalski's Cossacks of the steppes! Or better still, with the portrayal of honest, French-peasant life of Jules Bréton or François Millet. All these are realistic in one sense, and on the better side of realism—their foundation in real life—but life instinct with the expression of feeling and emotion. Their personages are actual, living, breathing human beings— not actors simulating them beneath the curtain of a stage.

In these—and especially with the works of the latter—

there is something more than Realism. The brush of a great artist, the magic wand of imagination and of genius, has touched and vivified the dull clod of humanity, and the soul of the man shines forth from amid the clay of its ordinary surroundings. The life of the toiling hind is faithfully portrayed. It is, indeed, the peasant, in his rude home or pursuing his usual avocations, but—taking him at his humble best, his moments of earnest endeavor, of aspiration and of adoration—they infuse the picture with that glow of true sentiment and feeling which can dignify and exalt the homeliest aspects of life.

Yet it is only fair to admit that certain realists, the pioneers of the school in French art, did a great work, a generation or so ago, in correcting the popular taste and breaking down a weak, false and conventional classicism. Such artists, for instance, as the historical painter Horace Vernet—clever, dashing and sensational. These swept away the old traditions, and gave opportunity for the cultivation of the most perfect technique that the world has ever seen. Considered solely as an art, without estimating its worth in the higher realms of Fine Art, painting, probably, was never brought to a greater perfection than it attains in France to-day.

Conspicuous amid this school for his strength, the very prince and apostle of realism in later days, the strongest and most satisfactory of all, we may cite Courbet, the Communist—he who, responsible for the overthrow of the Column Vendome, afterward paid some penalty for his vandalism. Master of the secrets of color—bold and vigorous both in interpretation and treatment—he commanded, says Jarves, "an introspective view into the primary elements of nature and of man, analogous to that exhibited in literature by Browning and

Walt Whitman." "Paint nothing that you have not seen! Show me an angel and I will paint you an angel!" —was the motto and expression of Courbet.

Wholly antagonistic in style was Corot—chief disciple and exponent of a more advanced dispensation. His is the great name in modern landscape art of that wonderful series which began half a century ago with Theodore Rousseau (inspired in the outset by the "naturalistic" Constable), and enumerates among its list of names painters of varying styles of treatment, though impelled by much the same principles in art, Lambinet, Daubigny, Diaz and Dupré. These are men of the same school, not as imitating one another, for each preserves his own individuality, but agreeing "in looking at nature not only for what she seems to the visual eye, but still more for what she suggests to the soul."

None of these noted artists were, after all, great landscapists in the sense of wide scope of subject and treatment, as Turner at least aspired to be. On the contrary, they are restricted to special aspects of nature and phases of scenery; so that only the wonderful mastery they exhibit and the charm with which they invest their special interpretations redeem them from the charge of monotony. Certainly were they working their vein in the realistic manner, we should tire of them in the extreme; but these wrought with thought, with deep sentiment and with loving feeling, and gave us the very poetry of landscape.

Rousseau revived the technical excellence of Ruysdael and of Cuyp, with a more natural and correct rendition of the greens in nature. Lambinet's pastorals, with somewhat less of poetic feeling, exhibit the same mastery of the resources of color, blended in most harmonious gradations, and are illuminated with sunlight, "painted as faithfully as pigments can represent it." The poetry of these two, exuberant with the joy of nature, is that of Robert Burns—of the early Tennyson (especially of Tennyson's "Brook")—and of the summer vision of Lowell's "Sir Launfal."

With Diaz and Corot, as with our American Wyant, it is the poetry of Bryant, and of Wordsworth; their sentiment being largely infused with a tender pathos, not to say a subdued melancholy, which reminds one of the author of "Thanatopsis," or of the "Intimations of Immortality." These together touch the extremes of the gamut of color: Corot with the light-greens of spring, and the silvery-grays of early dawn or twilight; Rousseau and Lambinet including all the affluent hues of summer; Diaz with the dark-greens and russet-browns of autumn.

Corot, it is admitted, is by far the greatest artist. The charm of his works consists not in their being mere transcripts of actual scenes—there is nothing of the photography of art in them. Either the artist penetrates deeper than many into the inner sense of Nature, or he imparts, like our own Francis Murphy, some quality of his own poetic imagination to the picture, which thereby gains the power to suggest and inspire moods of mind. Sweet mystery, dreamy reflection, tranquil enjoyment—these are states of mind induced by contemplation of the bewitching landscapes of Corot.

* * * * * * * * *

What is the true function of painting, what the province of the painter—and not of him alone, but of all artists and all art? Let us attempt to summarize, even though we should repeat. In so doing we shall by no means imagine that we are expressing any new thought, or one that has not been said in clearer and better phrase ofttimes before; yet in this intensely practical and realistic age, the reminder can scarce come too often. Once more let us put the old wine into new bottles!

It is then the province of art, not so much to represent nature as to interpret her. Nature, that is, in her highest; Nature at her best! The artist should have all the knowledge of technique which goes with the strongest Realism. He shall abide in that land for a season, but he may not inhabit it. He shall work through Realism into Idealism. He shall attain first to the body, and then to the soul that informs it.

The Poet first drinks at the fountain of preceding poets; he is an imitator before he is original. "He lisps in numbers ere the numbers come."

The Sculptor may well study first, and long, anatomy and models of classic beauty, till at last the flowing outlines of grace shall naturally and fitly drape the form whose face shall image the grand conceptions of beauty and purity that his artist soul shall shadow forth.

The Painter should, indeed, study nature. To him, all Nature and all Art should render up their secrets of light and shade, of form and coloring. Nature in sunshine and in storm: the broad prairie, the mountain cliff, the tumbling waterfall, the surge of ocean, the desert sand; the blue skies of Capri, the brassy glow of Egypt, the opal tints of Labrador;—all these should be known to the great painter. What then? Shall he stop at the pictured representation of these things on canvas? If so, what has he achieved? Simply a magnificent colored photograph!

No! He must, first of all, perceive what is picturesque in nature, what is worthy of translation, and then give us all this and far more than the form and tint of mountain, sea and sky. He must shed upon the canvas that glory without which, rock nor tree, nor curled wave, nor tinted cloud has valid excuse for being; that glory which, shining in the soul of the artist, an inner sense of something finer than all these, in a mystic world within or beyond, shall reflect upon the canvas before us, suggesting a yet greater glory:—

> "The light that never was on sea or land,
> The consecration and the poet's dream."

To enter into the finer sense of things around us; to follow out the suggestions of beauty and glory that ordinarily lie hidden in grass or flower, in tinkle of waterfall or tone of speech, in glow of sunset or tender irradiation of the face we love; all this appeals to a sense that, for want of better naming, let us term the poetry of life!

We choose to take it for granted that all the Finer Arts are correlated, and all pervaded in their higher forms with a spirit and essence, to grasp at whose expression we must reach far beyond all mere *representation* of things we see around us, however beautiful they may be. This spirit may find some manifestation alike through kindling eye, through eloquent or rhythmic speech, through music, through all the elevated forms of artistic expression:

> "The kindled marble's bust may wear
>   More Poetry upon its speaking brow,
> Than aught less than th' Homeric page may bear.
>   One noble stroke with a whole life may glow,
> Or deify the canvas till it shine
>   With beauty so surpassing all below,
> That they who kneel to idols so divine
>   Break no commandment, for high Heaven is there,
> Transfused—transfigurated."

THE OLD KNICK.

> "Stranger on the right
> Looking very sunny—
> Obviously reading
> Something rather funny.
> Now the smiles are thicker—
> Wonder what they mean!
> Faith, he's got the Knicker
> —bocker Magazine!"

So sang, once upon a time, John G. Saxe, the funny versifier, prince of doggerelists and poet-by-brevet! In that primitive period, "Riding on a Rail"-road, was actually something so novel as to be commemorated—and the early age of the railway was also the era of the Knickerbocker Magazine. With the multiplicity of periodicals now issued, there is such an embarrassment of riches, that one has only to choose—if he can—between them, and may only regret that he shall miss unavoidably many bright, enjoyable things, from sheer inability to read—and pay for—so many monthly magazines.

The generation of fifty years ago was not troubled in this way. For them, and for at least twenty years of their lives, the old "Knickerbocker" was the valued and only worthy representative of American Literature.

The writer is happily reminded of this ancient magazine, now unknown probably even by name to the present generation of readers, by the presence in his library of two stray volumes of its series—and these in turn remind him of a few early numbers that he discovered in his father's book-case, during his early boyhood, and of the delight with which he perused their contents.

We are not unmindful that there flourished, earlier or later during this period, various and sundry other monthlies; notably those located in Philadelphia, including Godey's Ladies' Book, Peterson's and Graham's Magazines. But these occupied a minor field, and however ambitious, could hardly claim to represent American literature. Godey's and Peterson's were devoted to the ladies—those of polite society—and were especially affected by young girls just graduated from the "female seminary" of that period, long anterior to the day of Vassar or the co-educational university. Each number of these periodicals started out with a highly-colored, lithographed "fashion-plate"—made up of wonderful, wasp-waisted, artificial divinities—and, next to this, as frontispiece, a "steel-plate" of almost equally impossible natural inanities, representing some child of earth, of the fairer sex, so highly idealized and etherealized as to present the strongest possible contrast to her sisters of the fashion-plate. The plane of every-day life, and the pencil of passably good drawing, were rarely attained in any of these artistic productions. Their letter-press was composed of sentimental verse, or stories by young writers, of such themes, for instance, as how the country mouse—an exemplary and sensible girl—went one winter to return a visit of the city mouse—her fashionable cousin—and got badly snubbed therein, but succeeded, nevertheless, in carrying off the spoil of the desirable and sensible city millionaire, who could appreciate goodness when he saw it!

But if this style of fiction was unexciting it was, at all events, unobjectionable. The Ouidas and Saltuses did not get into the magazines of those days. If insipidity prevailed, happily impurity was lacking. Godey was made up of pretty pictures, poetry and patterns, together with the "prunes and prisms" of prose, and Peterson's was like unto it, at five cents less per number. Graham's, as a "gentleman's magazine," professed a little more virility in its literature, and gave a trifle more originality.

So for a score of years, Knickerbocker was practically without a rival in its own field, that of literature which really possessed a literary quality. Starting in 1833 with Charles Fenno Hoffman, it came the next year into the editorship and part ownership of Louis Gaylord Clark, under which it continued throughout its history, and almost to its final close. In 1849, Harper's Magazine was started, but for several of its earlier years, that scarce came into competition with the Knickerbocker as a purveyor of American literature for, in the outset, the contents of Harper were largely "pirated" from the English periodicals. As some one smartly said at the time, its bill of fare showed that it "breakfasted on Thackeray, dined on Dickens and supped on Punch." With the advent of "Putnam's" in 1853 came an American Magazine, of high class—one that is hardly surpassed by the best of those published to-day—and comprising, in one periodical, almost all the choice features now included among the whole list of the present.

In the meantime, the good old Knickerbocker, for all those years, had at its command for contributors about all who gave dignity and honor to American literature; beginning with Hoffman, Cooper, Paulding, Irving, Halleck, Bryant, Miss Sedgwick and Miss Leslie, —of the old school—and including many of those of who have since come to the front—Longfellow, Holmes,

Hawthorn, Whittier, Aldrich—as well as still others, who, though "promising" in their day, are now forgotten wholly, or survive only in name, to prove how uncertain and brittle may be a literary reputation. As last year's leaves in the new spring-time, "the woods are full of them;" but alas, they lie dead upon the ground instead of fluttering green upon the boughs!

The contents of many volumes now classic in our literature first saw the light in Knickerbocker. Between its covers first appeared Irving's "Crayon Sketches," Longfellow's "Psalm of Life" and "The Skeleton in Armor," Ware's "Zenobia," and the "Tanglewood Tales" of Hawthorn.

Closely identified with the old Magazine is the memory of Willis Gaylord Clark, twin-brother of its editor. Willis was himself editor of the Philadelphia Gazette, but contributed frequently to Knickerbocker, notably a series of sketches entitled "Ollapodiana," which, in apparently desultory manner, happily mingled sparkling wit or genial humor with sentiment and pathos, and an occasional gem of poetry. These sketches added much to the early popularity of the Magazine—but their gifted author died young.

With all its talented contributors, however, its varied store of good reading in prose and verse, its success depended, after all, far more upon its editor's own efforts than is usually the case with a magazine. The real Knickerbocker was neither Washington Irving or his fellow contributors, nor any ideal old Dutchman in "knickerbockers," planted in an old, high-backed, carved chair on the title-page, but an actual, indefatigable, irrepressible Louis Gaylord Clark! Every month, his potent personality spake out cheerily and unmistakably in the last twenty or thirty pages, from the "Editor's Table," and in the "Gossip with readers and correspondents." His voice rang out clear with hearty, genial good-fellowship, often effused with rollicking, boisterous mirth, and then again, warmed into an eloquent or poetic fervor—or anon lapsed into tender cadences of pathos. What did these pages contain, or rather what did they not contain? They were a literary mélange of the first order; an Ollapodrida, a Salmagundi, and a Pot-Pourri of wit and of wisdom, of frolic and of fun! The touch was, perhaps, less delicate, the wit not so refined as that of his brother Willis, but it flowed forth an exhaustless stream of good things, from month to month, and from year to year.

The lines on which the old Knickerbocker Magazine was laid down, were much the same as many a goodly magazine craft has been built upon since. Their framers have somewhat closely followed the Clark model, or, where they vary from it, the divergence is more apparent than real. For instance, in Harper's Monthly the material of the old Knickerbocker "Editor's Table" has been taken apart, and reconstructed into an "Editor's Easy Chair," in which genial Curtis sits; an "Editor's Study," in which Howells handles lovingly every favorite volume; and an "Editor's Drawer," into which Charles Dudley Warner "puts in a thumb and pulls out a plum" or a few nectarines in season. The honest old Dutch table of the Knickerbocker had plenty of stuff in it to construct all this furniture out of—albeit the lumber was in a less finished and highly polished state than these moderns fashion it. One thing is true, however, of Louis Gaylord Clark; he was a "square man"—as square as his Editor's Table—and yet he could well "a round unvarnished tale deliver."

All good things come to an end however. After a continuous service of a quarter of a century, the veteran retired from the editorship. Already many of the brightest contributors to the magazine had left it, captured by and into the columns of some of its more stirring and aggressive rivals, of whom there were now several Richmonds in the field; including the popular "Harper," the staunch "Putnam," and the brilliant young "Atlantic." The strife and turmoil of the Civil War had come, and with it a new class and generation of readers. A new king had arisen who knew not Joseph. It were better for the old Knickerbocker, to whom all this stir and strife was uncongenial, to step down and out, rather than to have it said of him ungraciously and ungratefully,—
"Superfluous lags the veteran on the stage."

And still he lingered! One who has been a public's favorite, so hates to have the curtain rung down and the lights put out on him, for aye and all!

At last however, in his sixty-third volume, Old Knick was finally merged into another magazine, under the title of the "American Monthly Knickerbocker"—which, however, soon passed out of existence. It was, in a literary point of view, high time for the demise;
"For when his step grew feeble and his eye,
Dim with the mists of age, it was his time to die."

Still—in grateful memory of many a pleasant hour he gave in days lang syne—the writer is glad to say a good word even yet for the Old Knick.

PUTNAM'S MONTHLY.

Lovers of good literature of some thirty-five years ago that survive to the present, will recall with genuine pleasure the memory of the old Putnam's Magazine. In the above statement, one makes no account of the lapse of time nor of changing tastes; for surely, once to love good reading is never to lose its appreciation. In this case, so personal was the affection of its readers, that not only will the richly varied pages of the old magazine be fondly remembered, but also, in close association, the once familiar pea-green cover and the stalk of corn on either hand that framed the title thereon. That title-page cover came to be as well known as had long been the old Dutchman with his pipe and chair to the lovers of the "Knickerbocker."

It is the good fortune of the writer to possess a set of this periodical—both of the original series and of the "revived Putnam"—and these volumes now add the property of rarity to that of literary value, since outside of a few libraries they are scarce to be discovered.

This magazine was founded in January, 1853, by George P. Putnam & Co., a firm of book publishers noted as well for the uniform merit and high literary character of their publications as for their liberal treatment of authors—who, in turn, held the firm in grateful esteem, instead of distrusting them, in common with the whole race of publishers, as their "natural-born enemies."

The story of the origin of the Monthly was pleasantly

told in some gossipy letters long years afterwards, by two of its editors, Charles F. Briggs and George William Curtis, on occassion of the revival of the Magazine in 1868. The plans, it seems, were brought to light at a dinner-party given by the publishers, at which were present those immediately concerned and a few literary friends and to-be contributors. With Briggs and Curtis was associated also as an editor, Parke Godwin, the brilliant politcal writer and son-in-law of William Cullen Bryant.

Into the Magazine went not only the high hopes and ardent endeavors of the rising author of the popular "Nile Notes," but also all the bank notes of which he had become possessed thereby—and possibly the addition of a few of his own "notes of hand" as well, for his share of capital toward a part ownership—and, a few years afterward, when the firm went by the board, in the crash of '57, he had the misfortune to find his little all swept into the vortex of the liabilities of the concern.

But it was a brilliant junto of young writers that took upon themselves the burden of launching the new literary craft, assisted, indeed, by many of the ablest authors of the period, and attracting soon to their pages many yet unknown, but fresh and vigorous contributors. The general scope and plan of the periodical need hardly be outlined here, as it was essentially that of the "Atlantic" later on, of which, indeed, it was a brilliant precursor. Its reviews of current literature, both European and American, and of the progress of Music and of the Fine Arts—of which latter, indeed, there was then but little to chronicle—were most creditably handled by the editors, who also contributed most effectively to the body of the Magazine. Notable among these early articles was a series of papers which discussed from a high plane, the politics of the country and the policies of administrations, from the pen of Parke Godwin. Able, vigorous and fearless, these could scarce fail to make a deep impression upon the public mind—and what was still more needed then, the public conscience—especially as they were soon given a more permanent form by collection into a published volume. One of these broadly comprehensive yet trenchant articles, published in the October number of 1855, on "The Kansas Question," was particularly welcome to the young men from the North, settled here and then contending at great odds with the Slave Power.

Denounced by the Democracy as outlaws, harassed by Territorial governors, and proscribed by presidents, in the height of their discouragements they were grateful to know literature for their friend and the power of "Putnam's" on their side.

Another of the editors—that distinguished Mugwump, as well as genial "Easy Chair" in Harper's of to-day—then demonstrated his talented versatility by a dashing charge into the ranks of the plutocratic fashionable society of that era. His series of "Potiphar Papers," evinced him a humorist and satirist of high order—though showing somewhat the influence of Thackeray—and added early popularity to the Monthly. These, too, were soon reprinted in book form, followed, not long after, by another series of contrasting quiet and home-like sketches entitled "Prue and I."

And then, too, the Sparrowgrass papers of Frederick S. Cozzens! What lover of genial wit and rollicking humor in that day is going to forget the man who wrote, "It is a good thing to live in the country," and of the ludicrous haps and mishaps of that life! There has been a great development of "funny" Americans since then, whose wit is more exaggerated and broader, if not deeper—"the woods," and the syndicate "plated" newspapers are well *nigh* "full of them"—but in the ranks of genuine American humorists, and within the category of fun that, if not "fast and furious," is at least jocund and genial, we would still insist on reserving a good place for "Our American Cozzens."

There were not wanting either, genuine "sensations" to be exploited in the Magazine; one of the earliest of which was that of the problem of the lost (and found) Dauphin of France, suggested in that taking title of the first article of its series, "Have we a Bourbon among us?" It was a matter of genuine historic interest, and its discussion helped to swell the circulation of the Magazine.

Here too was broached, for the first time publicly in modern days, another controversy which, unlike, perhaps, that of the case of the Rev. Eleazar Williams, has never yet been finally set at rest, but like the restless ghost of Banquo, refuses to "down" for good at any authoritative bidding—the Bacon-Shakespeare controversy, started by Delia Bacon herself.

Here is the article—"William Shakespeare and his Plays, an inquiry concerning them"—the first page of January, 1856; published as a first instalment, with a note from the editors, commending it as "a bold, original, most ingenious and interesting speculation as to the real authorship of the Plays"—and as "the result of a long and conscientious investigation on the part of the learned and eloquent scholar, their author"—yet disclaiming, of course, any responsibility for such literary heresy.

It was a stone thrown into still water whose resulting circles have kept on spreading and widening; but this writer, having just re-read the article, would aver that the literary vigor, which went with the original cast, has never since been equaled by any latter-day Ignatius "Baconian." Beside the range of scholarship evidenced, there was an elevation of thought and feeling, a quality of literary style, and a power of imagination that have usually been conspicuous by their absence, in the later disquisitions on the subject.

On the other hand, one of the greatest of modern Shakesperians, Richard Grant White, first came into the arena of Shakesperian criticism, about the same period, in this magazine; of which papers therein, a book was made, entitled "Shakespeare's Scholar." Of all the commentators, Grant White stands as among the clearest, the most vigorous and the most logical—except, indeed, when occasionally lapsing into rhapsody, beguiled by that transcendental theory of the "unconscious possession" of "supreme, all-embracing," superhuman genius.

We might go on almost indefinitely, and cite many more notable books, made up in whole or in part from the pages of Old Putnam; wonderful papers by Lowell, Thoreau, Herman Melville, Tuckerman, Quincy, Clough! The stories were especially good—only the very best of those printed now-a-days comparing, indeed, with "Twice Married," "Israel Potter," "Miss Chester," "Stage Coach Stories," "Wensley," and dozens of others that might be named. "Of making many books" there was "no end," out of the magnificent literary material of the old magazine.

It was the rule of Putnam of the Old Series to give all its articles anonymously; a practice continued, in the outset, by the "Atlantic," but since abandoned by about all American periodicals. This had its advantages in that it gave the fairest possible show to new contributors, the merit of their articles not being overborne by their lack of "the magic of a name."

On the other hand, the famous writers were deprived to a large extent of the commercial value of theirs—the prestige that goes as an adjunct to recognized, demonstrated ability. It might seem that the reader too was deprived of some needed criterion, being thus left to his own judgment of what was best worth reading, unbiassed by any "sign-manual" of recognized authorship, and deprived of the satisfaction of his curiosity; but this was usually only for a time for—some way or other—such literary secrets were pretty sure to leak out, sooner or later, and in the meanwhile, perhaps, the speculations and "guesses" of the newspapers as to the authorship of a popular article, served to advertise the magazine more than the publication of the longest list of "noted" contributors.

In the first number appeared Longfellow's stirring lyric, "The Warden of the Cinque Ports," and while one knowing newspaper, assuming it as his, asserted that it showed signs of failing power, another found it to be undoubtedly but a weak imitation, from an inferior pen! So much for the infallible critics!

In a hurried glance over the volumes, one recognizes among the poems thus published anonymously—besides the lines on the death of the Duke of Wellington—"The Two Angels," "My Lost Youth," "Oliver Basselin," "Prometheus," and "Epimetheus," by Longfellow; "The Fount of Youth," "The Wind Harp," "Auf Wiedersehen," and others by Lowell; "The Conqueror's Grave," of Bryant; "The Ranger," of Whittier, and "My Mission" and "Young Love," by Bayard Taylor.

In 1857, the magazine, which had for some time previous fallen into the hands of Dix & Edwards, was finally sold and merged into Emerson's Magazine, a sickly periodical, with no part in the nervous vigor of intellectual life that belonged to Putnam, and then soon passed out of existence.

Eleven years after, when George P. Putnam & Sons had reëstablished themselves in the book-publishing business, it was thought safe to attempt the revival of the old "Putnam's"—once more under the editorship of Charles F. Briggs—and a determined effort was made to warm up the embers of interest that attached to the memory of the "Old Mag." The surviving contributors, who had once been so fondly attached, were appealed to; all of whom expressed the heartiest interest, while many gave substantial encouragement of renewed literary contributions—as also did a corps of new writers who gradually gathered around it.

Still, somehow, the old glories scarce came back. The young "Atlantic" now occupied very much the same field and cultivated it vigorously, and perhaps the business management was not such as to ensure financial success. At all events, after a respectable but not brilliant career of three years, the magazine was again sold out—this time to the Scribners—and on its ruins arose the successful "Scribner's Magazine," afterward renamed the "Century."

The old "Putnam's" has gone for good, but its old-time readers will still remember it gratefully. At a time when the Harpers and other publishers were fast bound by the chains of commercial self-interest to "Old Hunker" subserviency to the South and to slavery, Putnam's gave itself freely to the cause of free soil and free thought. Amid the arid desert of an uninspired, commonplace literature, parched by the dreary drought of dough-faceism and famished through a dearth of faith, it was a fountain of sweet waters welling up refreshingly; it was the green palm waving in an oasis and casting a cooling shade, grateful as "the shadow of a great rock in a weary land."

To one fresh from perusal of the current number—August, 1890—of the Atlantic, it might hardly seem that the transition from the magazine of one-third of a century ago were, after all, a violent one.

Here is a poem by Whittier—"Haverhill"—that, at eighty-three years, betrays no loss; and here too, the "Autocrat," with whom we breakfasted in the first number of the Atlantic, still entertains us as brightly and genially as ever, "Over his Teacups!" Though every one enumerates "The Autocrat of the Breakfast Table," among his list of "One Hundred Favorite Books," yet even its covers scarce held anything racier than this poem of "The Broomstick Train."

Remembering the mellow wine, so clear and fine, poured out as a libation at once to the youth which passes and the youth that endures, by Longfellow in his "Morituri Salutamis;" recalling Bryant's "Flood of Years," and all the later verse of Lowell, Whittier and Holmes, produced after fifty—yea, sixty years—it were scarce too much to affirm of these venerated authors "the best wine is the last."

THE GOLDEN AGE.

'Tis told in Hesiod's ancient rhyme—
  And still we love the mythic story,
How earth had once a primal time,
  Its sun, a more transcendent glory.

The whole world at one altar knelt,
  And man to man as brother,
In tranquil peace and concord dwelt
  As children of one mother.

That Golden Age on earth once shared,
  When angels walked with men,
Poet and seer have long declared
  Some day, shall come again.

But waiting not, let each and all
  Restore some truth to fable olden,
Some bliss of Eden ere the Fall,—
  Bring back, with love, the Age that's Golden!

1855 TO 1854, GREETING!

(At a Reunion of Kansas Pioneers.)

As ONE who, later born, yet envies not
   The earliest in primogeniture,
But holds in honor and affection sure
That elder brother, whose the happy lot
To heir the crown for which they jointly fought,
   Then freely shared with all beneath the sun
   The heritage of Freedom kept and won;—
So we but honor, what yourselves have wrought.
First, highest place, we gladly you assign
   Who earliest strove, 'mid darkest storm and stress:
'Tis haply yours to know, ere life decline,
   The chance Fate gave into your hands to bless.
Founders and fathers of a mighty State,
We hail you as of all most fortunate!

THE LOUNGER.

AT HOME AND ABOUT.

Our minds travel when our bodies are forced to stay at home. —*Emerson.*

TO THE POET OF THE PARK.

THOU, whom the early hasteners to toil
Discern 'mid shadowed copses, as they pass,
Noting each glint of pearl on dewy grass,
Inhaling all the fragrance of the soil—
Welcoming the echo of wood-robin's note,
(Sweetest of all that burst from feathered throat);
Who markest with delight each opening bud,
And each new leaf that trembles in the wood—
No shy thing animate but trusts thy mood,
And loves thee, as thou lovest bird and flower:—
Here may we find thy haunt at later hour
Of drowsy noon, stretched out by foot of tree,
Its shadowing leaves thy curtain-canopy;
Hearing of thousand insect wings the drone and whir,
The wren's sharp twitter and the hum of bee:
Thou chartered favorite! 'tis given thee
To feel each tingling pulse of Nature stir,
To be in touch and unison with her!
* * * * * * * *
But as the sun rides down to glowing west,
And length'ning shadows stretch athwart the glade—
Leave thou to multitude thy Park, displayed
In garish light, and evening dress arrayed:—
Seek thou that scene which Nature claims her best,
And climb with me the slope of Oread's crest!

Of tree and shrub a varied range explore!—
Thy birds thou shalt not miss, but gain the more:
And more of all!—Thine eye, in pensive mood,
May range o'er stream and valley, plain and wood,
Viewed thus from far, a peopled solitude—
Where storm-clouds darken—or where sunshine smiles
Above a circling arc of four-score miles!
In quiet noon, in far-off leagues of skies,
Becalmed at sea, float snowy argosies;—
At silent eve, from distance infinite
Of dim horizon trembling into sight,
Rolls toward this cliff of shore—toward you and me,
A misty blue of darkling, heaving sea!
* * * * * * * *
Thou, who canst love a Nature vast and grand,
Come where the heavens and earth alike expand,
Come and behold this ocean of the land!

TRAVELS AT HOME.

I.

THE writer, who, in his own time and earlier days, disported somewhat as a rambler, has now settled down, in his green old age, into a confirmed, professional Lounger, vibrating between the chimney corner in winter and the street corner in summer time—indulging in an occasional stroll, it may be, as far as Bismarck Grove, or even the borders of the Wakarusa, in the pleasant spring or mellow autumn days.

It occurs to him that there may be those who, like himself, or for other reasons than age or indolence restricted in their wanderings, but unrestrained as to their loiterings and ponderings, might yet be able, with a trifle of assistance or direction, to glean a goodly harvest of visual enjoyment within their circumscribed area of observation—especially when that range is so prolific in beautiful scenic effects as is the vicinity of our town of Lawrence. Should this last statement be received by any with incredulity, and a smile at the local egotism of the Lounger, let it be his pleasing task to attempt the conversion of such to a share of his belief in its truth.

There are always two classes of people in this world—those who believe that the rainbow touches the earth very far away from the beholder, and those who never fail to see it arching and glowing almost immediately over their own heads. Of the two, the latter class is probably the happier.

Again, the division may be made on other lines, which are happily far from inclusive. First, those who travel far and see little. Second, those who see and learn a great deal without stirring any distance from home. Of course, there are home-keeping folks in plenty, who see little and learn less ("home-keeping folks have ever homely wits," says the adage of such), while fortunately, on the other hand, we have some travelers whose many angles of *incidents* are fully equaled by their angles of reflection; which, indeed, should always be as true in foreign travel as in Physics.

But were we confined to the first two classes alone, the Lounger would hardly chose him who had "traveled the wide world all over," and yet had brought back little of value save a few diamonds and dress-suits, that he had saved duty on. Of such, the trunks are generally better filled than their heads, and they fairly merit the cynic observation once falsely fathered on Humboldt, and as falsely applied to a noted American—that he "had traveled farthest and seen the least of any man he had ever met."

While the Lounger might desire for himself the wider range of observation, he grants the meed of his humble admiration to those who, tethered by a short rope, have closely and exhaustively cropped the field of knowledge within their reach; or to put it in more æsthetic phrase, who, from closest sympathy with nature, have taken into their hearts and minds every charm and secret which she discloses only to her chosen votaries. To such, an island may be almost as comprehensive as a universe!

Gilbert White, of Selborne, found the little parish he inhabited a microcosm of England and the world—and in making its natural history noted, created a classic of

English literature. Thoreau could find almost everything in flora and fauna around the shores of Walden Pond, which itself, like Wordsworth's Rydal Mere, is only a mill-dam in extent. A Pennsylvanian—Dr. William Darlington—publishing a local botany of his native county, made his "Flora Cestrica" so complete and comprehensive that it became a standard, securing its author the association of leading naturalists the world over; his own name given to a family of plants in remote California, and his bust placed near that of Sir Joseph Hooker in the Royal Gardens of Kew.

These are but a few instances of the many that might be cited as to the capacity of the born naturalist to garner rich harvests from limited fields, and the rule holds good, to a great extent, with respect to the beauties and scenic effects of nature, as well. One need not traverse sea and land to find them; to him who is by true sympathy instructed in their mysteries, they will be "here and there and everywhere" around him. He can behold as richly tinted skies in Kansas as in Italy; more glowing reflections in the despised Kaw than the Arno at Pisa or Florence; as broad and beautiful a landscape spread out from the summit of Mt. Oread as from the heights of far famed Fiesole, albeit not so classic.

* * * * * * * * *

All very well, says the reader, but, dropping sentiment and coming down to business, whereabouts do you propose to begin your "travels at home?" Gentle reader—gentle or simple, whoever you be—the Lounger does not propose to start you off at all at the end of this long exordium! No! take breath for a week first, and in the meantime, any day in the week, go and stand at the intersection of Massachusetts and Winthrop streets—

in the center thereof—and look east, west, north and south! And if right there, where hundreds pass and repass each other daily, you can discern naught of beauty at the end of any of the green vistas, the Lounger has grave doubt whether he ever wishes to take you along with him at all.

II.

The candid reader (and all my readers are of the candid kind, for none others will care to scan these papers) will frankly admit that the Lounger did not seize him perforce and rush him off hurriedly upon these Travels at Home. Every journey presupposes a certain amount of preparation given beforehand, and certain requisites of travel laid in. Now, for our shorter ramblings, about the only requirement the Lounger would insist upon is a receptive state of mind. This is absolutely essential to the proper appreciation and enjoyment of the scenery, and will be found just the happy mean—as far removed from any gushing tendency on the one hand, as from a *nil admirari* spirit on the other. It is the tendency of young and unsophisticated travelers of ardent and impulsive nature, to exaggerate and "gush." It is the fault of experienced ones to be hypercritical—persistently unwilling to discover anything to admire. Even when you direct their attention to some lovely scene, they can only impair its charms by belittling comparison with their reminiscences.

Which prefer you, gentle reader, the Caviler or the Gusher? As for the Lounger, he commits himself to neither company, but like the colored gentleman to whom was presented the alternative of two terrible roads of theologic dilemma—"dis darkey takes to de woods."

* * * * * * * *

More than a week ago the Lounger left his reader "planted" at the corner of Massachusetts and Winthrop streets. Apologizing for the incivility, if he will now climb with us the stairways of the National Bank building, he will be amply rewarded for the fatigue by the fine views he will obtain from its upper windows, embracing nearly all the city and much of the beautiful country surrounding. From the north windows, the view embraces the dam, the mills and bridges, the river and its valley; with North Lawrence and Bismarck, and the beautiful rolling bluffs beyond, which, in the days when the old tribe inhabited them, the early settlers were wont to call the Delaware Hills.

These finely rounded bluffs, ranging in from the westward almost parallel with the river's course, and then trending off to the northeast toward Leavenworth in smooth and graceful promontories, are always a beautiful element in the landscape around Lawrence, whether clothed in the white snows of winter, or, as now, in emerald verdure of spring. Especially are they in their scenic glory on those days of fitful sky, when the fickle sun, shining between shifting clouds, flecks them with alternate light and dark, as sunshine and shadow in play chase each other over their fair surface. This effect in the distance is most beautifully exhibited however, from the summit of Mt. Oread.

Descending from this favorable near-by post of observation, we take our way to the wagon-bridge. This is the favorite haunt, not only of occasional amateur artist for sketching, but of the whole tribe of Loungers. It is their "custom always in the afternoon"—especially of a Sunday afternoon in fine weather, to resort here in flocks and swarms. It is the Ultima-Thule in wandering of their shoals—as the dam beneath is that of the shoals of cat-fish in spring time.

There is always something attractive to your true Lounger in the sight and sound of running water. It soothes and satisfies his soul. Like Tam O'Shanter's witch, "a running stream he dares not cross"—that requires too much effort—he just stays on the bridge and watches it hurrying by. Here, with the music of the water rushing against the piers, and the roar of its torrent dashing over the fall, he surveys the finely curved shores and the still reach of the waters above, dotted with an occasional pleasure boat and "white sail gliding down," or the even stretch they take, churned into foam and breakers, as they rush on eastward under the railroad bridge; the little island in the stream below giving pleasing variety to its career. Here, too, the professional Lounger watches with interest what current of life flows past. He is almost willing that the dam should "go out" once more—that he might be furnished with the mental occupation of seeing it rebuilt! Here, again, from his comfortable perch, he can watch the Santa Fé trains as they come and go; though alas! he misses, of late years, their taking on of passengers from the platform below! His crowd of old, on those occasions, was wont to be so miscellaneous in composition that one was reminded of Tennyson's prelude to "Lady Godiva:"

"I waited for the train at Coventry,
I hung with grooms and porters on the bridge."

* * * * * * * *

Referring to the future a visit to the charming green glades of Bismarck Grove, we retrace our steps to our starting point.

Here we are tempted by the verdure of South Park to

take that route, from whence, on the way, at the crossing of Henry street, we could obtain a fine foreshortened glimpse of West Lawrence, with the western bluffs for a background—as the Rocky Mountains wall up the western end of the streets of Denver. But, instead, we turn to the right, and follow Winthrop, to the corner of Tennessee. Here, with the beautiful Trenton-red of the corner residence as a foreground, we get a fine bit of effect, looking north to the river and beyond.

From the head, each of Tennessee and Ohio Streets also, pleasant river views are to be obtained. It was from the latter point especially that the regatta on the Kaw, a few years ago, was seen to the best advantage.

The vistas to the south along both these avenues are favorable specimens of street views in Lawrence, which at this season of the year, from the wealth of foliage framing them, are a perpetual delight to a lover of the greens in nature.

III.

An Old Dramatist—thus easily the Lounger evades an issue of vexed controversy—an old dramatist represents to us on one occasion, the boon companions of a graceless and reckless reprobate, receiving reports on his sad condition; sick unto death, out of his mind, and babbling of green fields! The worn out old voluptuary, the wonted familiar of the stones of Eastcheap street and tavern, is now stretched on a bed of pain from which he shall never rise. A reflex from the innocent and happy days of boyhood sweeps across the jangled chords of his unconscious brain, and "he babbles of green fields!"

They have no hopes of him now: that lapse into such senseless vagary of the imagination—such strange freak of that tongue which never wagged to them but of worldliness and wickedness—denotes that he is very far gone, indeed! Forsooth, "he babbles of green fields!"

* * * * * * * *

One day in Brussels, the Lounger, in descending Rue Montagne de la Cour—that thoroughfare which pitches so steeply down from the plateau of the New Town to the Old—chanced to glance through a break in the row of tall buildings that line it on the right; and, looking athwart broken lines of red and gray tiled roofs, over mellowed masses of buildings which slope downward to the plain, the vision stretched past the old city, its pinnacles and spires, and on across river, field and forest, to a far off horizon.

It was a glance as through a window just opened in a high tower, while yet our feet touched the solid earth. It was a magnificent picture, deeply framed in by massive walls. Its unpremeditated, yet wondrously picturesque effect, was such that the Lounger may scarce lose its vivid impression so long as memory shall last.

Now the Lounger did well to pause and enjoy this wonderful picture—for it was his "by right of discovery," not being discoursed of in any guide book—and then, too, he was entitled to all he could get, in part payment for thousands of miles of toilsome journey. Why should one go abroad unless to see something?

But the busy Bruxellians, thronging past by the thousand every hour—why should it be anything to them? Suppose one should delay his companion with: "Hold here, a moment! Just look at that fine bit of effect through the opening there!" His comrade would reasonably exclaim impatiently: "Oh fudge! Don't stop mooning here in the way of people! you can see that any day of your life. Come along; we have only five minutes left to reach the Bourse and place that order for stock, and, after that, you know we agreed to meet those ladies, to lunch at the '*Milles Colonnes.*' If fine scenery is what you want, come with me to Switzerland this summer!"

* * * * * * * *

Another day abroad, the Lounger was strolling listlessly along the Heeren Gracht, near the heart of that "northern Venice," Amsterdam. If you have any doubt as to the identity of the Lounger, you can always detect him by the slowness of his gait. On this occasion he was on his way to a picture gallery, and consequently strolling even more tardily than usual. It was a pleasant day and a peaceful scene. Here were no hurrying crowds, and he had chance to loiter and enjoy the beautiful studies of color; the water of the winding canals, with gray stone bridges, and green trees bordering, all contrasting with the differing but harmonious tints of red in brick of wall and tile of roof, afforded by the quaint old buildings. It was all artistically perfect in tone of color, and shockingly "out of drawing," so far as lines were concerned—the Lounger being charmed not to find a single straight one in the whole picture. The streets wound in and out, and the buildings were guiltless of verticals anywhere. They give the stranger the impression, at first, that they are out on a "jolly drunk"—but that is not the matter. Too much water is the trouble instead of anything stronger.

Well, the Lounger would have enjoyed these views greatly, but for lack of company. Do you know how hard it is to laugh when alone? Just try it, my humorously inclined reader! You will find that it requires something extraordinarily funny to constrain you to laugh all by yourself. You need sympathy also, to enjoy scenery to the utmost. Through some occult suggestion, the Lounger's memory flew back to a day spent in Bismarck Grove, when he was associated with two ladies of culture and artistic taste, as "hanging committee" of the Art exhibit. "Now," thought he, "if Mrs. Gray and Mrs. Black were only here, how much they would enjoy all this."

* * * * * * * *

Well, just the other day, another lady—who is sufficiently near the Lounger to serve as mental stimulus, and preserve his wits from perishing through inanition—was at a tea party—what they call now-a-days "afternoon luncheon"—and sat at the same table with our two good friends, Mrs. Black and Mrs. Gray. Breaking one of those pauses with which the time spent at five-o'clock-teas is generally made up, jolly Mrs. Gray exclaims:— "Who is this Lounger that is mooning so much in the Journal of late—do you know? I should think he might amend his style a little, as well as find something worth writing about, beside perpetually harping on the beauty of the verdure around here. Who is he anyway—do you know?"

The question was so direct, and withal so embarrassing, that the lady addressed—our Queen Consort—who passes by another sobriquet in society, had to truthfully acknowledge that *she* was "the Lounger," herself.

—Now, unkind and unappreciative Mrs. Gray—and to be unappreciative of the Lounger's articles surely is unkind—you should know an author better! It is impossible for him to "get over his stile," and almost equally difficult for him, at this beautiful season of the year, to "keep off the grass," no matter how many warning signs you may put up.

The Lounger can well conceive that this "babbling of green fields" must seem strange to many, and absurd that he should stand at street corners, admiring green vistas, when he might better go into the shops and transact some business, thus helping to "make business" these dull times! Or, indeed, that he should affect to climb the stairways of any bank building in Lawrence for other purpose than to get his note extended therein!

Nevertheless, so long as people will love the beautiful, and all cannot spend time and money to seek it in travel in foreign lands, let the Lounger plead once more that our own folks may keep their eyes open to that beauty which is all around them. On any one of these fine days, come up to the summit of our Pisgah—Mount Oread— and see the Promised Land! Men from other shores have found the landscape admirable in comparison—and it is well worth gazing upon, even if it is our own! As the minister sometimes says—"let me repeat:"

> "We look too high for things close by—
> For far-off joys—and praise them,
> Whilst flowers as sweet bloom at our feet,
> If we'd but stoop to raise them."

ON MOUNT OREAD.

A JUVENILE member of the Lounger's family, once, on occasion of discussion as to a debatable visit to be made, attempted in all sincerity "to move the previous question" on the subject, by the following expression: "Why, I *must go, I'm invited!*"

If invitations do imply an imperative necessity of corresponding attendance, the Lounger will certainly be on hand at the Kansas State University, on Commencement week—for he has been invited! Yea, through the unerring certainty of the mails, and the generous favor of the "powers that be," he has received no less than three copies of a printed invitation to be present. Gratefully acknowledging the compliment—in triplicate—we accept all three. We are coming in full force. And we shall kindly encourage our numerous friends to come and share with us all the enjoyments that pervade this happy season, especially the intellectual feast "set down on the bills" of the Commencement *menu*. We shall go through the whole "bill of fare," as the countryman is said to do sometimes at an unwonted high-class hotel. The delicately flavored "purées" and the appetizing "consommés" being already disposed of, we shall now follow on with all the solid viands and "sweets," in due successive course: on to the valedictory and the benediction, when—the parting word of good cheer having been spoken—"the favored guests" of the occasion shall be kindly sped to the world outside; but certainly not without a "souvenir" each, in the shape of "sheepskin!" In all the after years—among all the jocund feasts they may share, at all the "groaning boards" they may surround—haply there is "one class" of these guests of to-day that may never receive thereat any "favor" which shall be esteemed more precious than this!

In the intervals between the sessions (if any are left), those of us that come up to this Jerusalem, or Mecca of ours—the University—only on occasion of this annual pilgrimage, will do well to improve the opportunity to see what has been added since last year. Verily "the world does move," and our University with it! Once, as the Lounger easily recalls, only a little building at the north slope of Mt. Oread; now with its five commodious edifices—and one of these, magnificent Snow Hall, which arouses the enthusiasm of all beholders! And yet, these "surface indications" are but faint suggestions of what is going on inside. "For particulars, see"—Annual Catalogue.

We will visit the library in its changed quarters—a new edition of itself, revised and improved.* We shall not fail to view the exhibit in the Drawing and Painting Department. It is "all set down in black and white" this year—or rather it is hung up in those tones. When we have enough of these comely moderns, we will call upon the language departments, for their classic models of Art—not forgetting the statues and casts of Demosthenes, Kikero, Mercury, Zeus,—and the rest of "them literary fellows" of ancient times! But chiefly our researches will be made in Natural History, among the birds and beasts; our far-away cousins (like unto our-

---
*Now embracing 13,000 volumes, instead of number given on page 97.

selves—and yet happily unlike, even when skeletonized): the giant Megatherium, the Pleiosaurus, the Pleonasm, the Pterodactyl, the Dactyls and Spondees. Possibly the Lounger has got some animals mixed in here that really belong to another department. But at all events, he is very sure of the "bulls and bears," and especially of Prof. Dyche's buffalo. Poor fellow; there used to be no mistake about him! There was plenty of him then, and he spoke (or rather bellowed) for himself; but now:

"Last of his race, on battle plain,
His voice shall ne'er be heard again."

* * * * * * * *

Verily, Mr. Chancellor, this castle of yours "hath a pleasant seat." What university in all the land hath such an outlook; one embracing such a magnificent scope of country on every hand? The Lounger fancies he hears some one whisper "Cornell;" but this deponent, knowing naught of Ithaca, "saith not."

Our venerable poet, Holmes, said not long ago—contrasting in reminiscence the outlook from the Harvard of his earlier years with that now circumscribed on every hand by intrusive, neighboring brick walls—that it was a rare good fortune to a boy to be born and reared where he could have the prospect of a natural horizon. If this be so; if the daily contemplation, in early years, of pleasing and inspiring scenery will have its specific effect both inward and outward—in developing a love for the beauty and freedom of nature, and in a corresponding widening of the mental horizon—then the students of Kansas University are especially blessed in their opportunity. Certainly the natural horizon before them is wide enough to suggest and inspire mental "breadth of view."

The similitude of a morning landscape—looking eastward—to the aspect of human life as seen from the stand-point of youth, is no doubt sufficiently trite, and yet it often comes upon the Lounger with renewed significance, as suggested by the view from this noble hilltop. Fresh and dewy, sparkling yet distinct is the foreground, as its waves of verdure swell upward to his feet. Beyond, the landscape spreads out fair to his vision, without shadow of cloud upon its face; but soon its middle distance merges in the haze that lies over the valley, concealing all but the tops of intervening ridges— the dim landmarks that youth intends to make upon its journey.

Farther on, the broad horizon can scarce be even faintly traced, lost in the effulgent sunlight; the glamor with which hope irradiates the bright future of youth!

There is no perspective to this picture; the glow of faith has effaced it; youth needs none!

And confident Manhood: it, too, scarce feels the want of "distance." Its landscape is symbolized by the view from the south and west windows of the University, just before noontide. A view of fertile fields and meadows to be tilled; and beyond, of stream and, if need be, of hills to be crossed with easy endeavor. Near to us— beautiful slopes, graceful and noble "lines of descent," that make and mark the transition from hill to valley. How beautiful it all is; how practicable everything is; how easily accomplished! Everything is to our hand in this view; everything is possible. We reach forth our hand, and lo, it is done! Life, health, strength are ours; Nature, jocund Nature, herself is ours!

"'Tis in life's noontide she is nearest seen.
Her wreath of summer flowers, her robe of summer green."

And then comes the afternoon of life. The prospect is no longer all sunshine: shadows, clouds, and sometimes the darkness of storm sweep over it; but afterward comes the "clearing up," with all the charms and joys of nature enhanced by the contrast. The channels of experience are deepened, the springs of life renewedly filled, even by the storms that pass over us. Life has more significance. The tones of the afternoon landscape are more varied, richer and deeper; the tints of nature more harmonious—if we look eastward. There is enough perspective now; the horizon begins to grow distinct and sometimes sharply defined!

*   *   *   *   *   *   *   *   *

There is a seat that often finds the accustomed Lounger now at eventide. It is one of the top ones of those cyclopean steps which buttress the entrance stone stairway of Snow Hall. Before him, the tops of the trees in North Valley of the campus, jut up from the tangled depths of wild-wood below. Beyond, lies a beautiful little valley, and the pleasant homes of West Lawrence. Farther on, a little circlet of water gleams within the foliage that lines the bank of the river, of which this apparent lakelet is but one of the windings made visible. Beyond this again, and foreshortened, lie the everlasting hills, which have seen differing races of savage men come and depart. Behind them, as the light fades from the hill-tops, its glow is caught up by the mountain-like banks of cloud that lie piled in massive cumuli; which glow, and then fade in turn, in the dusk of even.

And lastly comes the night with its majesty of the heavens, wherein, one by one, come out "the stars invisible by day."

IN THE WOODS.

> "Of all the beautiful pictures
> That hang on Memory's wall,
> The one of the dim old forest
> Seemeth the best of all."

SOME one has suggested that the Lounger's peregrinations have been quite restricted in their range, even for Travels at Home: embracing no greater extent than from the river to Mount Oread! Admitting the fact, is not that doing pretty well for a lounger? Walter Scott said of his "Marmion:"—

> "Mine is a tale of Flodden Field,
> And not a history:"

and the Lounger might plead that his is chiefly a tale of the town-site, and not a gazetteer or map of Kansas!

He is free to admit that thus far he has had no occasion in "these presents" to utilize the railways of his country, but it is, perchance, high time to promote at least the patronage of the Lawrence livery stables. Let us then go abroad—say three miles or so out of town.

* * * * * * * * *

On his recent visit here, Colonel Higginson was most impressed, as he more than once expressed, by the wonderful change made in the appearance of these prairies by the growing of trees. Thirty-two years had elapsed since he had seen Kansas, and the transformation of the landscape as exhibited from Mount Oread was marvelous in his eyes. Not so much, after all, for "the improvements," the numberless houses, the tilled fields, and other signs of cultivation (for these he had somewhat anticipated), but for the trees that aid so charmingly in diversifying the landscape, and which change what was once a monotonous expanse of plain, to a scene of sylvan as well as of pastoral beauty, scarce to be excelled.

On the other hand, the old-time and primitive forests have largely disappeared. These, it is true, were chiefly confined to the borders of the larger streams, but just here at Lawrence we had quite a large body of timber, above and below the town, on the south banks, and especially across the river, where North Lawrence now stands. Of this, Bismarck Grove, with its fine old spreading elms, is but a remnant, haply spared. That, as the Lounger recalls, about marks the confines of the woods where they touched the prairie, but all intervening was a heavy, though not dense growth of timber, embracing grand old oaks, walnuts, cottonwoods and sycamores, through which the first road to Leavenworth wound deviously after leaving the ferry. In those days, when returning from an occasional visit to that town, belated at nightfall, after traversing the Delaware Reserve, how interminable and darkling seemed the path, meandering between those hoary monarchs, until we reached at last the sandy shores of the river, from whence, across the flood, shone dimly out "the lights of home." Then, routing out from his cabin on the bank, the old French half-breed ferryman, we were soon set across by skiff, or "floating scow," and back to the welcome precincts of the "historic city." But—the Lounger is here being betrayed, he fears, into that garrulous reminiscence which indicates approaching senility.

* * * * * * * *

IN THE WOODS.

From the slopes of Mount Oread, looking a little to the north of eastward, there is still to be discerned what appears quite a heavy body of forest, embracing the timber of the river about the mouth of Mud Creek, and stretching back to the upland beyond—a fair fragment of that heavy growth which once marked the whole course of the river as seen from this point.

But the Lounger is "minded" to take his "gentle reader" with him on the little voyage of discovery which, within the short distance of three miles, may afford an intimate impression of a bit of forest, or, at least, a study for a "wood interior." This is reached by taking the middle Lecompton road, west from town. After climbing the steep hill westward (or rather south-west) of "Hillhome," we turn in our seat to take a retrospective view of Lawrence. From this point, only the western and northern portion of the city is visible—the vicinity of the post-office being especially prominent—but the view embraces much beside that is very attractive; including, in the near foreground, the river valley, which assumes quite a park-like appearance, as beautified with the rounded, deep masses of foliage that line the little winding stream or "branch" that intersects it. Lawrence lies in the middle distance, and beyond it the broad belt of forest already mentioned as extending to the eastward, while just to the right, Eudora is seen sleeping on the billowy plain that sweeps on to the far horizon, dim in the blue and hazy distance.

It is but a short step after leaving this "coigne of vantage," till our road dips down into a wooded hollow; at first but a narrow gulch, but soon widening into a broad ravine, which, expanding, seeks the lower level of the valley of another Mud Creek. Oh! hapless soil of rich, fatty, or ashen black, which so appropriately bestows the unpoetic name of Mud Creek upon so many Kansas brooks—would we ever willingly exchange your homely productiveness, to gain the sparkling limpidity which correlates with the sterile granite of mountain streams? Is there no happy medium; no land where the timely and bounteous rains of heaven—falling on the just as well as the unjust—may descend on fairly fertile alluvium, without carrying in solution to the streams and seaward so much of turbid yellowness and blackness!

To the left and onward, as we descend, trends a range of picturesquely sloping and rolling hills, covered with woods that late in the season afford quite a beautiful effect of variegated autumn coloring. Here is about as good sketching-ground for foliage—that is foliage in mass—as the Lounger is acquainted with in this vicinity. Then too, as the road winds along the margin of this forest, one comes in contact with sights and scents—of plant and shrub, of leaf and blossom—which carry him back with the swift telegraphy of memory to those days of boyhood when all such were very near and dear to the fresh and opening senses, the avenues to the mind and heart of youth.

Traversing this road in the early summer, one gets at almost every step a luscious whiff, the scent of the wild grape in blossom. It is well worth the ride from town— and the carriage hire—to inhale once more this delicious fragrance, and to hear echoing in the deep wildwood the sweet notes of birds, especially the sweetest and clearest of all, that of the remembered "wood-robin" of the Lounger's boyhood—which our naturalists insist should be known instead as the wood-thrush. The Lounger loves, indeed, the "dim old forest" of the present, not only for its own sights and scents and sounds, its "pictures" of to-day—but for the hundreds of others it suggests—the never-to-be-forgotten recollections of childhood:

> And still in memory fresh as then
> I seek each thicket, glade and glen,
> Where woodsy odors wild and sweet
> Rise up at every crush of feet;
> Where waves the plumy fern, and dank
> Green mosses carpet rock and bank.
> On knolls that boast "the Barrens" name
> The mountain-pink, a sheet of flame,
> In distance burns—but glowing near,
> Azalea's trumpets fill the air,
> With pungent perfume blown afar.
>
> The kalmias waxen clusters spread
> On rocky slopes—while overhead
> The dogwood drops its petal snows,
> And fragrant with each wind that blows,
> By roadside blooms the sweet-brier rose.
> * * * * * * *
> Far down along the forest glades,
> Upspringing, mid the woodland shades
> With graceful, true and tapering lines,
> As California's sugar-pines—
> The Liriodendron skyward showers
> A thousand glorious tulip flowers.
>
> Tinted with orange, green and gold,
> Its cups a honeyed nectar hold,
> Where bee and humming-bird in tune
> Make glad the lightsome air of June.
> Each cup, amid the glistening leaves,
> A largess to the summer gives,
> For dews of heaven it receives.
> —Queen of all forests yet, to me,
> The Pennsylvania tulip-tree!
> * * * * * * *

IN THE WOODS.

Nor one of all the thousand rills
Amid the everlasting hills,
Dashing from rock to rock their spray,
Or stealing silently away;
From Ammonoosuc's windings shy—
To Mercede's sources far and high
Where sharp Sierras pierce the sky;
Not one, or all of these, whose praise
Poets sing in tuneful lays,
Shall quicken pulse of mine in joy,
Like that one brook I knew as boy.

The rill that all the livelong day
With rocks and pebbles smooth at play,
Made everlasting roundelay.
Where oft I paddled "barefoot" feet,
Built my mill and sailed my fleet,
Just where the woods and meadows meet.
On sweeter stream I ne'er shall look
Than one little nameless brook,
Whose springs of life were near to mine,
—The brook that ran to Brandywine!

IN AULD LANG SYNE.

THE Lounger has been revisiting the pleasant haunts of his youth—in Chester County, Pennsylvania. It may be from the predilection of early association, but—after some extended journeyings in later years—he still returns to these scenes with the fancy that none elsewhere are fairer or sweeter.

His headquarters from which to take varied excursions, are made at the old county-seat town of West-Chester. Would that the Lounger could make this charming little town known adequately to the world; but that is, humanly speaking, impossible. West-Chester is *sui generis*, difficult to describe, and scarce to be compared. However, one might intimate, to mind of a New Englander, somewhat of its perfections by saying that, in a manner, it is a kindly-sedate and scientific-minded Pennsylvania-Quaker Northampton. This would convey but a limited notion of its characteristics; as might also the further statement that it has long possessed the finest collections of minerals and mortgage-bonds, and has always used the best microscopes and made the best ice-cream of any town in the Union. What more could heart—of any town—desire! And yet some restless people—and there are always such in every community—were scarce satisfied even with this; and but lately would fain try the desperate experiment of a transfusion of "new blood"—of "enterprise," we think they called it—into its body politic and economic. They even went so far as to get up a "Board of Trade" and get out an advertising book of "advantages." What was peculiarly Quakerish about this, however, was that it actually confined itself very closely to the truth!

Had this rash experiment of transmogrifying it into a "manufacturing" town succeeded, it would have badly spoiled it for a Lounger. But fortunately, the healthy old borough survived the shock, and now goes on still in its good old ways of quiet improvement, while its people continue to live well and live long—as before.

—With excellent roads throughout, charming drives may be taken in every direction from West Chester: along upland slopes, winding through shaded woods, or dipping down into the valleys of streams: through lush meadows, fragrant with the scent of mint, where the clear brooks run, sparkling and tinkling over their beds of pebbles. We retrace all these again with ever new delight! Four miles to the southward takes one to the valley of the Brandywine, at the point famed as the scene of its Revolutionary struggle. On the heights of Birmingham, overlooking the valley, still stands the old "Meeting-House" of dark bluish-gray stone, around which the fiercest combat centered, and in whose grave-yard adjoining, many of those who fell were buried, almost indiscriminately. Near by, Lafayette was wounded, and when the American lines were here broken, Washington, at Chads-ford, was flanked and compelled to retreat, uncovering Philadelphia to the possession of the British.

On the round slope's crest, by the walnut trees,
Whose boughs have rustled in every breeze,
Through a hundred years of storm and calm—
Stands the Meeting-House of Birmingham.

By its grave-yard wall, with bayonets set,
Stood our sires of yore—and the foeman met:
O'er the trampled turf, with carnage wet,
Rode Sullivan, Greene and Lafayette.

Now the long-fled years have left small trace
Of their strife and blood, on that battle-place:
While the Quaker farmer, silent and calm,
Still holds his "meeting" at Birmingham.

* * * * * * * *

Crossing the Brandywine, and retracing to the westward for a few miles, the previous march that morning, of the columns of Cornwallis and Knyphausen, brings us to Longwood, near the site of Old Kennett Meeting-House. Longwood is almost the only home and assembly-place of the Progressive Friends. This sect—"if sect indeed that might be called which creed has none" (to misquote Milton)—was a "liberal" offshoot from the Quakers, during the later Abolition period, and though quite limited in numbers, for awhile made "quite a noise in the world;" embracing among its speakers at its annual reunions, many noted and gifted Reformers. The building is but a modest frame structure, while just across the road is Longwood Burying-Ground, now best known as the last resting-place of Bayard Taylor.

The scenery of this section, while amply deserving, has been fortunate in securing three poets for its singers: two of them, Buchanan Read and Bayard Taylor, belonging to it by birth, and the third, John G. Whittier, by long-time association with its people. Among the latter were the venerable Friendly pair whom he has celebrated in his "Golden Wedding of Longwood."

"Fair falls on Kennett's pleasant vales and Longwood's bowery ways,
The mellow sunset of your lives, friends of my early days."

The mastery of the portrayer of scenery, the born "poet of places," is evidenced by the manner in which Whittier sees and seizes the salient characteristics of a section, and thereby pictures the whole in a very few words:

"Again before me with your names fair Chester's landscape comes,
 Its meadows, woods and ample barns, and quaint stone-builded homes:
 The smooth-shorn vales, the wheaten slopes, the boscage green and soft,
 Of which their poet sings so well from towered Cedarcroft."

But none sings it better than Whittier himself. In these four lines he misses little else than the clear-flowing streams, which, indeed, are scarce distinctive—being shared as well by New York and New England.

Bayard Taylor's grave is marked only by a plain low granite column, relieved by a medallion portrait in bronze. On one side lies the love of his boyhood, his first wife, Mary Agnew, whom he married upon her death-bed. On the other, is buried his brother, Fred Taylor, Colonel of the First Pennsylvania (Bucktail) Regiment, who fell, bravely leading it on the first day of the fight at Gettysburg. On his monument, no less than four tributes are rendered by admiring poet-friends—R. H. Stoddard, George H. Boker, Phebe Cary, and his brother. Bayard himself can well afford to sleep there without further memorial on his tomb than the simple inscription thereon: "Being dead, he yet speaketh;"—for many loving brother poets have embalmed his memory in verse, whose lines will live in literature. Among these comes to memory those of Whittier, addressing the neighboring scene:

"Oh vale of Chester trod by him so oft—
 Green as thy June turf keep his memory! Let
 Nor wood, nor dell, nor storied stream forget,
 Nor winds that blow round lonely Cedarcroft!"

* * * * * * * *

And so we make next, a short pilgrimage to "lonely," "towered Cedarcroft;" the home which Bayard Taylor made in the days of his prosperity, returning to dwell amid the scenes of his boyhood. The road we follow, traverses many of the scenes he describes in his own "Story of Kennett."

The house stands about a mile north of the village of Kennett Square, a station on the Phila. & Baltimore Central Railway. Bayard's only child, his daughter Lillian, has married in Germany; her mother also has returned to her native German land, and Cedarcroft has passed into the ownership of an eminent Philadelphia surgeon, now retired from practice. Evidently the doctor's ideas of "a fine place" are nearer the conventional ones than were those of Bayard Taylor. The native wildness of the fine woods which the poet loved, has been sacrificed. Some of his "immemorial chestnuts," which "westward" stood "a mount of shade," have been cut into rails; the underbrush in front has been cleared off, and the noble oaks trimmed up, so as to afford a vista of the mansion from the road.

The old farm gate at the entrance, has given place to a beautiful "porter's lodge." The house, a somewhat irregular pile of brick, two stories in height, with its "ivy-mantled tower," is, however, apparently much as Bayard had it, in the days he entertained therein right royally—too hospitably indeed, for his purse—his friends and fellow poets.

* * * * * * * *

On the slope of a hill, three miles north of Cedarcroft, still stands an old two-story brick school-house. In this building, known as Unionville Academy, Bayard Taylor received, after the district school, all the education afforded him within walls. Its principal—Jonathan Gause, an old Quaker—was, as the Lounger recalls him, an almost perfect type of the school-master in Goldsmith's "Deserted Village," in his ponderous dignity and severity, tempered withal by that enthusiasm for learning which seldom fails to arouse the latent ambition of the scholar.

In the lower story of the academy building, a district school was kept; and there, as an urchin of tender years, the Lounger first made acquaintance with Bayard Taylor, then a youthful pedagogue of seventeen. This teaching was but for a term that bridged over an interval between his own schooling and going to learn the trade of printer.

Two years passed by, wherein the writer had graduated one step—or rather a whole flight of stairs—upward into the Academy, where Bayard had studied. One day the Lounger, whilst at the principal's desk, by special favor, reciting his *Telemaque*, found his lesson interrupted, as a fresh-faced stripling came in to bid his old preceptor "good-bye." It was Bayard Taylor! With many misgivings, his father had bought off his "time" as apprentice at the printing office, and now he was off to see Europe, "With Knapsack and Staff!" In those days, and among that steady-going people, the tour of Europe meant far more than a journey around the world now-a-days, and such an undertaking for young Taylor, with his means—or rather with his lack of means—seemed, to the Quakers of Marlborough and Kennett, in the highest degree Quixotic. In a moment he had quietly passed out of the school-room, and into the region—in our youthful imaginations—where dwelt unexplored mystery and romance.

Another two years—and our old academy is all excited by a visit from Bayard Taylor. His letters of travel—

compiled into a volume entitled "Views Afoot"—have become very popular, and kindly favored by the literary world, from the freshness of their observation and the novelty of their point of view. The young American who, making the tour of Europe on foot, has written so pleasantly of his experiences, is the literary sensation of the day. In our eyes, he has achieved fame, for he has published a book! From this on, the world is familiar with the literary and personal career of the poet-traveler. In the old Academy, the teachers would sometimes take the new scholar to a special section of one of the long pine desks, lift up the lid showing the initials carved on the inside thereof, and say: "I give you the seat of Bayard Taylor!"

It is not within the Lounger's present scope—which is simply one of reminiscence—to render any estimate of Taylor's character and abilities. Poet, traveler, novelist, biographer, lecturer, editor and critic—so truly versatile as to be eminent in many fields—time will determine, ere long, his true place in literature. Minister of the United States to one of the great empires of the world; his country's honored representative in that land where once, a poor stranger youth, he toiled on foot from town to town; and, finally, his obsequies thronged by great ones of the earth;—the writer would praise and honor Bayard Taylor to-day, in recollection, not chiefly for his talents and achievements—worthy as they were—but for one thing, than which his life held nothing worthier or manlier. It was that in the day of his assured success, when culture had graced and fame had crowned him, he forgot not to love and cherish the humble country girl, the companion of his youth, who had given him in those early days her maiden love and faith. The pages of his

"Poet's Journal" still speak eloquently his life-time tender memories of the early loved and lost.

* * * * * * * * *

One of the features of this section, in primitive days and down to the Lounger's boyhood, was the prevalence of the old-time country inn or "tavern." This prevailed, indeed, from the old Colonial period down to the advent of the railroad, which rendered obsolete the old thoroughfare of travel—the turnpike road with its Conestoga wagon.

These ancient hostelries, devoted in those days to the "entertainment of man and beast," survived their legitimate use, deteriorating to the entertainment of man only in such manner as tended to confound him with the "beast." Briefly, they survived simply as drinking places, where the rural population were too often tempted into habits of loafing and tippling. This was their dark page of history. Now they are fast passing away; an enlightened public sentiment seeing no occasion for licensing them longer. Their venerable, tall, framed sign-posts, with the "sign" swinging and creaking in the wind, have been taken down. These signs, with their appellations, were curiously reminiscent of old England, as commemorated by Dickens:—The Red Lion—the Black Bear—the White Horse—the Ship—the Anvil—the Sickle and Sheaf—the Hammer and Trowel! The similitude of these things, as painted on the sign by the local artist, was often something wonderful to behold!

As reminiscent, too, of "Merrie England," as is the scenery of this section, are the old hawthorn hedges which prevailed here as in no other portion of the United States. These are fast being supplanted by Osage-orange; and only occasionally now may you find some ancient ones, grown almost into trees, where you may sit "beneath the hawthorn's shade."

The local names here—especially those of counties and townships—were largely "brought over" by Penn and his companions, from their old homes:—Berks, Bucks, York, Lancaster and Chester being names of prominent shires in England. The emigrant of all times has been glad to remind himself in every way possible of the land he left behind. "The skies change but not the man."

—The Lounger is no exception to this feeling:

> To me, though wand'ring East or West,
> Where Nature spreads her choicest, best,
> No mounts a fairer prospect show
> Than thy north fields, East Marlboro';
> Whence Bradford towns and Laurel woods,
> And Newlin's meadows, wet with floods,
> But heighten th' opposing scene,
> Where plains of Fallowfield lie green,
> With Doe Run Valley spread between;
> And westward rise, like sloping lawn,
> The hills of Highland and of Caln!
>
> How oft in boyhood's early day,
> I viewed those hills ten miles away,
> And longed for all the world unknown
> That lay beyond their purple zone!
> * * * * * * * *
> That world unknown has come to me
> From Eastern hills to Western sea.
> In manhood sought, th' horizon shifts,—
> Its purple glamour fades—and lifts.
> Onward!—the glamour lifts, and fades,—
> Till age draws on with twilight shades.
>
> Haply if, when no more for me,
> Earth's glamour rests on land or sea,
> To eye of faith, the glory lies
> On world unknown beyond the skies.

BY THE SEA.

> "Spirit that breathest through my lattice—thou
> That coolest the twilight of the sultry day,—
> Gratefully flows thy freshness 'round my brow;
> Thou hast been out upon the deep at play,
> Riding all day the rough blue waves, till now,
> Roughening their crests and scattering high their spray,
> And swelling the white sail—I welcome thee
> To the scorched land, thou wanderer of the sea!"

FAMILIAR to memory since boyhood, these lines with which Bryant opens his address "To the Evening Wind," came fondly to the lips of the Lounger, one recent evening, as the train which had borne him over weary miles of "scorched land" swept into the cool stone paved station of the West Jersey Railroad at Cape May. The "evening wind" was duly on hand to meet and greet us, together with the "smiling host," to whom we had telegraphed for rooms. In spite of Shenstone's famous saying, and though it is no doubt true that in point of importance the inns are justly one half the "inns and outs of travel,"—at this season of "heated terms," the *summum bonum* of felicity consists in finding the welcome warm, but the rooms cool.

And the supper-room too, was delightfully cool, as the descriminating waiter—with the possible anticipation of a future fee—placed us most favorably before an open window looking directly out upon the ocean, and so near that the spray could almost dash in upon us, laden with that salted scent that is so deliciously fresh when approached from the fevered land. Once more the sights and sounds and scents of the unforgotten sea are ours! The rich hues of evening are deepening on the waters—blue waves come chasing each other shoreward, combing there into translucent green, and breaking into lines of foam. Farther out, the sea is picturesquely dotted with sailing craft; some with sails filled with the evening breeze, glistening in snowy white; others more distant, glimmering in ghostly gray—and yet others, hull down, with masts only peering above the horizon, telling freshly the tale of earth's convexity. It is a scene for a William T. Richards water-color—and indeed the visitor to the Metropolitan Museum can note in the series of Richards' sketches there exhibited, that he has reproduced nearly every possible peaceful effect of sea and shore and sky to be witnessed on this Jersey coast, which he has long frequented.

With all the æsthetic conditions so favorable, including the hotel table itself, the Lounger was enabled to "materialize" a very fair supper, and the precedent then and there established has since been so consistently pursued, that there is now every prospect of his realizing an additional pound for every five dollars spent—which would be a very fair rate of exchange indeed, for a seaside season.

* * * * * * * *

With many visits since intervened, the Lounger is always freshly reminded here of his first trip to Cape May, made a score or two of years ago, in his early days of diffidence—in his green and callow youth. This—as is well known—is an old-time resort, and was then a famous and fashionable watering-place. As a boy on his father's farm, he had heard of its charms, and determined to "take in" a few of its enjoyments, in the interregnum "betwixt the walnuts and the wine"—the wheat and oats harvests. He had, indeed, to spare but a few days and a few dollars, which latter would not reach very far at any of the long, tall-pillared caravanseries which then lined its shore, in the fashion that continues to the present.

But the little glimpse of watering-place life there afforded him, showed in dazzling and bewildering colors to the neophyte. There were hops in the dining-room, after it was cleared, in the evenings, and the unacquainted Lounger stood at the open door as if it were the gate of Paradise—longing, but not venturing in—even though a gilded youth from the city, whose acquaintance he had made on the boat, kindly but patronizingly invited him to "join in." (The Lounger now shrewdly guesses this gay young fashionable to have been a clerk from some Market street grocery.)

Most unwillingly, however, he left this enchanted ground for home—first paying, of course, his bill at the office. That could hardly have been a long one, so soon; neither was it "as wide as a church door, but it was enough." It was deep enough to reach almost to the bottom of the pocket of the young man from the farm: but after all, he did not "mind" that half so much as the careless remark with which that genial but gorgeous clerk in the office sped the parting guest: "You dont stay long with us, Mr. Lounger!"

Long years have flown since then, and that clerk has possibly ere this "disremembered" his casual, and perhaps, politely meant remark; but you see the Lounger has not! With some murmured reply about urgent business requiring his presence in his city counting-house, he turned his back upon that celestial scene—with that vivid (and it seemed to him unfeeling) sentence ringing in his ears: "You dont stay long with us, Mr. Lounger!"

* * * * * * * *

Alack! Cape May is still here with its long colonnaded hotels, its bands of music, its hops and its throngs—the latter not so much increased since other resorts have multiplied—but the youthful glamour is all off with the Lounger! These people are not princes and princesses, nor even lords and ladies in disguise. They are simply good, common-place, every-day folk taking their recreation—dipping and tumbling around in the surf with quite unromantic *abandon* in the mornings, and lounging in the halls after dinner, with paper novels hugged to their bosoms (as Howell notes), the fore-finger still holding the spot where the sensation flagged.

And these giddy girls in tennis suits, "having a good time" in their own "careless and happy" fashion;—are these the beautiful fairies, the "angels without wings" of the Lounger's early vision? It would seem not—and yet possibly they closely resemble their mothers and grandmothers,—and it is only "the grave stranger come to see the play place of his" boyhood, that has changed after all!

One thing he will scarce dispute,—that people seem to be taking their summer holiday in a comfortable, and on the whole, sensible fashion here at Cape May. Either because the tide of extreme fashion has deserted this for newer resorts, or for some other reason, there appears to be less show and more substance in the enjoyment here. This is now, at least, a good, sensible, comfortable place for those who come to the seaside for recreation in rest.

And whatever other changes the years may have wrought and brought, the ocean is here just the same as ever—still rolling its tireless waves on this noblest stretch of beach on the Atlantic coast; here with its freshening inspiring breezes, and its thousand charms of changing sky, reflected by morn, noon and eve in changing sea. Even with its grotesque groups in the foreground, rolling like so many porpoises in the breakers, it refuses to be vulgarized. An old-time Lounger, whose point of view may have changed on many things, can be thankful that nature here is still unchanged for him! That—

> "The radiant beauty shed abroad,
> On all the glorious works of God,
> Shows freshly to his sobered eye
> As e'er it did in days gone by."

IN THE SURF.

IN THE year 1609, Henry Hudson, a distinguished English mariner, on his third voyage to this country, attempted to enter the Delaware and subsequently landed at what is now known as Cape May, after he had narrowly escaped losing his gallant galliot, the "Half Moon," by ship-wreck. Fourteen years later, Cornelius Jacobson Mey, a Dutch navigator, rounded the south point of New Jersey, and named it after himself. He called what is now the Delaware, South Bay, and the mouth of the Hudson, North Bay; hence we still have the name North River for that stream which empties into the latter. The first European proprietors obtained this part of New Jersey from nine Indian chiefs by actual purchase. During the 18th century, the Cape attracted the attention of fishermen who were engaged in capturing whale, blackfish and sea-lions which then abounded in her waters.

The above is history, and is largely drawn, as is most of the historical and statistical matter of the Lounger's essays, from the local guide-book. Whenever you find him unusually replete with facts, you may set it down that he has been reinforcing his own splended memory of events that transpired in the last century, with some acqui sitions from the guide-book. It is altogether the safest way.

Notwithstanding, however, that the above cited is history, it does not inevitably follow that it may not be true. The Lounger would carefully guard the mind of the young reader against the impression that whatever is set down in the books as history is necessarily false. A strong presumption in favor of the authenticity of the above narrative lies in the fact that it does not pretend to claim that Captain John Smith discovered Cape May. The Lounger believes that this point is the only one along the Atlantic coast which this veracious and ubiquitous traveler failed to touch upon, either in his vessel or his Narrative. We are referring now, of course, to the original story-telling John Smith, and not to his multitudinous namesake of the New York city directory. Even the latter's "funeral knell," if tolled, would be too long a story altogether.

About ninety years age (to be exact) an old Quaker farmer, living inland a few miles from Cape May, used to come down to the shore to bathe in the surf. The neighbors regarded this as a very strange and rather dangerous freak withal. Even large vessels had been known to go to pieces in the breakers—and why not this crank? They used to follow him down on the chance of seeing this happen. Some of them doubtless were wreckers or wreck-savers by profession, and ready to claim the pieces, or salvage. They used to assemble thus on the beach to the number of twenty or more— say twenty-three to be exact—of a morning, on such occasions. Finding that the man came out all right and that it "would wash," they gradually took to imitating him. Hence the origin of surf-bathing!

The above is tradition. Tradition is not always as unreliable as history, but generally runs it very close. The difference, as the Lounger apprehends it, is about thus.—With history, either the circumstance took place in time and manner, and with the person indicated, or else it is absolutely false. With tradition, the *circumstance* probably transpired, or something like it; but, perhaps, a thousand miles away, a thousand years before, to a thousand other fellows. So it may be strictly true in a race sense, though slightly incorrect in an individual one.

Now, like the William Tell myth, this tradition of the origin of surf-bathing may not be properly located in Jersey at all, but may have happened in Sweden or Syria, to some of our Aryan or Unitarian ancestors. But they tell it down at Cape May all the same.

Anyway the custom has grown and multiplied until now-a-days we, who are of the "interior department," no longer "go down into the sea in ships," but in bathing dresses! Some of these are not exactly in "ship-shape" either. The whole Atlantic coast is being parceled out into bathing and watering-place stations—"cities by the sea." Every few miles, already you come upon them anew, sown thick "as leaves in Vallambrosa," or as empty tin cans around a western town on the plains. In a few years more Whittier might find no spot left where he could pitch his "Tent Upon the Beach."

—Were the Lounger an artist—a Clays, an Achenbach, a Richards or DeHaas, for instance—the world would soon be the richer for his sojourn by the sea. What magnificent marines he would then paint! Even if only a skilled amateur, he would still be attempting to imitate Rehn, or Bricher, or Niccoll, in transfixing upon paper some hint of the wondrous effects that enchant his eyes; translating some of the thousand harmonies of tint—the rich colors of water—into appropriate water-colors. As it is, he is only an "impressionist," with but little faculty of reproducing his impressions.

EVENING AT CAPE MAY.

I.

Rich gleams of gold upon a western sky,—
   Broad stretch of gold upon a quiet sea;—
Shore-seeking waves that softly break and die
   In flaky foam that melts upon the lea:
   Dieth the Day as soft and tranquilly!
The stars and crescent moon come out on high;
   The beacons on the Sea-Wall shine,—and far
   Across the bay, Henlopen's ruddy star
Kindles, and signals eve's departing light:
Slowly, on land and sea, descends the Night!

II.

A leaden sky hangs o'er a leaden sea;
   Keen lightnings quiver over wave and land;
Deep thunders roar,—and thunders ceaselessly
   A sullen surf, hard beating on the sand;
Before the gale the gull drifts aimlessly;—
   The tides of rain and surge meet on the strand;
—A murk of storm blots out Henlopen light,
And all the world fast darkens into night.

AT ST. AUGUSTINE.

On San Augustin's moss-grown wall
The tides of ocean rise and fall:
As lapping of the tides, Time sees
The course of empires, dynasties:
They rise, they fall, and who shall say
Save Time, who knoweth yesterday,
To-day, and shall to-morrow know,
Whether the ceaseless ebb and flow
Shall bear our Nation's fortunes on
To "heights of glory yet unwon"—
Or, late or soon, the Right defied,
Shall crumble every mount of pride,
And whelm her in Oblivion's tide!

Here throng the memories of her reign,
Once monarch of the land and main,
Queen of two hemispheres, proud Spain!
The centuries have come and gone,
Claiming her conquests one by one.
For all her deeds of valor done,
Here on our shores beneath the sun,
One massive fort of mouldering stone
Preserves her memory—one alone,—
San Marco, now Fort Marion.

*　　*　　*　　*　　*　　*

Without command or tap of drum,
Phantoms in armor, rayless, dumb,
How Fancy's shadowy legions come!
It needs not captain, troop and gun
To give the Old Fort garrison.

Lone figure from thronged History's page;
Last watch-tower of the Middle Age;
An outpost of a force withdrawn;
A lingerer, who waits alone,
Unconscious of his comrades gone;
A sentinel with ward to keep,
Who slept the centuries' dreamless sleep,--
Still standing thus, so silent, grim,
As Sleep and Death were one to him:
'Twixt waters blue and meadows green,
Thus stands thy Fort—St. Augustine!

INNS AND OUTS OF EUROPE.

When books of travel are full of inns, atmosphere and motion, they are as good as any novel.
—*Augustine Birrell.*

BEFORE THE CURTAIN.

Too early for the play! Musing I sit
'Mid hollow silence and a half-lit gloom,
While shadowy ushers move about the room,
Seating the early comer as they flit.
Unreal all the aspects of the place
Save one, whereon what pictured fair might seem,
A southern land of mountain and of stream,
Some artist bold the curtain fain would trace.
What though the painter crude in color be!
Now while the music swells into a tune,
Heard once in far-off land 'neath summer moon,
I blend his scene with hues of memory.
So, mellowing hour and throbbing tune and he
Draw back my heart, fair Italy, to thee!

* * * * * * * * *

Brightest of visions yet to me—
A scene of panoramic glory—
Of emerald hills and summits hoary,
Bold curving shore and promontory—
The last fond glimpse of Italy.
Thou comest back, Lake Maggiore!

How sparkle in the sun thy waves,
Or lap the shining sands so stilly!
What tender glow of color bathes
Thy far-off towns—or nearer, laves
The wall of tower and campanile!

The blue of distant mountain range
Whose summits to the clouds are given,
Makes symphony, in subtlest change,
With blue of lake and blue of heaven.

BEFORE THE CURTAIN.

On Maggiore's tinted tide,
His oars a stalwart boatman plied,
While, sheltered by the skiff's rude awning,
A lounging tourist stretched beside.

In light and shade, a glorious bay
Encircled by the mountains, lay
The lake—reflecting shores of splendor
Through glowing hours of summer day.

Isola Bella's terraced bowers
Slope backward from its clustered towers:
The gem of the Borromean islands,
Ablaze with tropic trees and flowers.

As eastward in the bay he drew,
The farther mountains lift in view,—
The five-cleft peak of Monte Rosa
And other Alpine heights he knew—

Whose names the boatman fain would tell—
The Fahlhorn and the Mischabel;
Whilst bold in front, Sasso del Ferro
Guards lake and lovely shores full well.

But ere the traveler seeks the zone
Of glaciered-granite Mont Leone,
The wind-swept realm of dashing torrent,
And frozen giants that guard Simplon.

Linger—and boatman, drop the oar—
Where Intra's prospect lies before;
Float idly by Pallanza's shore!—
For scarce, in galleries of the mind
Shall Memory's pictures clear be dimmed
Of vistas fair as e'er were limned
By artist's brush—in poet's story;
Of beauteous shores that curve and trend,
Of pearly peaks that heavenward tend—
Of skies that arch and downward bend—
(At rose of dawn or gold of even
When flushes all the face of heaven)
Reflecting all their radiant glory
In thy fair face—Lake Maggiore!

Up goes the curtain! As its folds arise
My vision fades—the music sinks, and dies.
Italia's sunny skies I view no more,
But moonlit ramparts of cold Elsinore:
A northern air with mystery is rife
And all the portents of the tragic life.

What is the real? Not life's usual pace—
This slow procession of the commonplace!
Let Life move faster with a varied train!
Vanish the Present! Give us back again
That fuller life of passion, joy and pain;
That pulsing life replete, come player, bring!
With mad unrest of Prince and guilt of King!

Such is the Drama's—such the Player's power:
His fancy thralls us for one magic hour.
This is the actual—this, that stirs and moves
To the soul's depths—these deepest hates and loves
That sway the heart to joy, the hand to strife!
Aught less is dream—this is the real life!

Oh power of Genius! While unreal seems
My real past as is the land of dreams—
Thou peoplest all the living world for me
With forms that never were, but aye shall be!

MINE EASE IN MINE INN.

> "Who-e'er has traveled life's dull round,
> Where'er his stages may have been;
> May sigh to think he still has found
> The warmest welcome at an inn."

THIS is not the first time, by any means that Shenstone's cynic lines have been quoted. They have just been brought freshly to the Lounger's recollection, through being cited anew in an artistic circular, charmingly illustrated, descriptive of the Ponce de Leon hotel of St. Augustine, which is said to be the finest hostelry in the world.

It would be a great pleasure to the Lounger to respond to this "card of invitation," and renew his acquaintance with Florida and that delightfully quaint old town, one of the oldest on this continent—through the medium of a visit at the Ponce de Leon, the Cordova and the Alcazar;—or, at least one of those "Spanish-Moresque Palaces, set amidst the luxuriance of the orange, the palm and the vine,"—with their "courts, plazas, marbles, mosaics, fountains, etc.,"—all "Spanish of the Renaissance period." As all these palaces are now under one ownership, it makes but little difference to Mr. Flagler which invitation the Lounger may accept—while "the best is none too good" for him. Yet inasmuch as, unfortunately, the "warmest welcome" of the Ponce de Leon, costs twenty-five dollars *per diem*—with no manner of welcome, of any mercurial degree whatsoever, on hand at less figure than eight dollars—the Lounger shall perforce have to stay where the temperature is a trifle cooler; at least until he can market his corn at over sixteen cents per bushel. Some writer once remarked that "Home, sweet home" was a good place, "however homely"—or words to that effect!

And yet, the Lounger, without going all the way with Shenstone, has some rather pleasant associations with hotels. After all, when you are far away from home, on a foreign soil—hungry, tired and sleepy after a long day's tramp or travel—the homely, country inn, or even the city hotel may prove a passable substitute for a home—on a pinch—and, at all events, beats staying out doors all night by a great majority.

For instance—take that dingy and dreary old Queen's Hotel in London! Somebody had recommended it to us when we went abroad, so we sojourned there for a season, and might have been much more miserable. If you are going to spend much time in the world's metropolis, the Lounger would advise you to take two hotels. Not both at once; that prescription might too much deplete your system—of finance. Quarter yourself, for part of your stay, at one of the caravanseries in the vicinity of Trafalgar Square, and you will be in convenient proximity and striking distance of many of the most interesting and enjoyable sights of modern London; the picture galleries, Regent Street, Pall Mall, "Green Pastures and Piccadilly," and a dozen others; as well as the more historic Parliament Houses, and Westminster Abbey.

But after you have "loafed and invited your soul" to your heart's content with these, then move down, bag and baggage, to the heart of Old London, the East City!

And there, for "Headquarters in the saddle"—you may find the quaint old Queen's Hotel as good as any. It is by no means fine—in a modern American sense, you may say it is hardly comfortable—but it is a part of, and in harmony with its surroundings; it shares in that atmosphere of the Past, which still lingers and envelopes the region in which it abides. It is, indeed, rather stuffy and contracted—it is "neither as deep as a well nor as wide as a church door, but it will suffice;"—especially if, like the Lounger, you have some toleration, yea, even affection, for things old and staid and set in their ways— good, stiff, old-fashioned ways, which, British though they be, were once those of our grandfathers' grandfathers— that we have outgrown and discarded long since. So if you put up there, you will end by putting up with them.

Geographically, the Lounger will locate the "Queen's" for you, by stating that it is immediately opposite the General Post-office, in St. Martin's le Grand, and directly between "Angels" on the one hand and "Bull and Mouth" Street on the other. There you have its situation, and who would wish a better! It is full within the glad sound of Bow Bells;—on your right, St. Paul's is scarce distant a stone's-throw;—farther away, to the left, is Smithfield, about whose "market," dead beeves now hang where heretic martyrs once burned;—just east of you is Guildhall;—down in front, Cheapside will lead you on to the Bank of England and all the money-changers of the world; while if you prefer to "take the back track," Newgate Street will land you at its prison, or at least give you a sharp turn into the Old Bailey!

. Extend the radius but a little, and you reach Blackfriars, the Temple, Fleet Street, Holborn, the Charter-House, Finnsbury;—or London Bridge, Billingsgate, the Tower,

Eastcheap, Whitechapel, indeed! You are within the historic bounds of the old City Wall,—you have the "freedom of the city," and may go where you list, provided you can thread the crowd by day, and take a policeman for safety by night.

Inside the old inn everything is awkward and inconvenient until you get accustomed to things not where you would naturally expect, but somewhere else. There is the dark statue of a man-in-armor that greets you in the hallway as you enter! There is the railed and partitioned "bar" in the center, just as Dickens describes it; with its liquids and its solids; its spirits and ale and stout; its cold joints and cheeses and dishes; with many odd and out-of-the-way things, kept there because they would seem to have no other and more appropriate place. And therein stands the prim old-maid bar-maid, who will answer your questions civilly, and give you information in crisp, bitten-off sentences, or tell you, more frequently, "I dont know sir, I'm sure." Also she will hand you out your letters when they come, and sell you paper and envelopes to write your answers—for "they never give nothing for nothing" in any English hotel.

And then, you may stumble up the steep, narrow staircase, with the edges all worn off by the wear of long years, to the coffee-room above;—and there you will be waited upon by the silent Will-yum,—the waiter, not the historic Him of Orange—who will wait upon you with more than the habitual reserve and slowness of the British waiter; and to all your questions he will answer, "I dont know sir, I'm sure."

"Remote, unfriended, melancholy, slow;" what was the matter with Will-yum? "Remote," he generally was— at the other end of the room when wanted,—slow he was and preternaturally slow,—but why apparently so "unfriended and melancholy," we never knew! Perchance, like the "oyster" he so much resembled, he may have been "crossed in love;" but more likely it was that they had stopped off his beer.

In the little, stuffy parlor alongside, we make the acquaintance of almost our only fellow-guests that frequent it—an old couple perfectly in keeping with the inn itself. The kindly, chatty old lady tells us that they are from Banbury Cross, that famous home of the "bun," and the haunt of that other old lady of nursery-rhyme, who mounted thereat a "white horse," long ere the "red-headed girl" had ever discovered and become associated with one in history. Her heart warmed to us Americans, because they had a son over here on the same continent; and becoming confidential, she told us that she and her spouse had been coming up to London annually, for a visit, for the past thirty years, and they always stopped there at the "Queen's." That her husband was much respected at Banbury, and was withal quite a scholar, which she was not. That he had written much for the papers, and *once had a letter in the Times*. "It were summat about 'the land,' or the Malt Tax" she believed, and "he do think the country have been injured bad by the way they do go on in Parliament." Although far from being her admired husband's equal in learning, she confessed that she had brought him quite "a nice little pot of money," when they married, which had given him quite a good start in the world. "When I was a slip of a lass," she continued—and so forth, and so on—in charming reminiscence! And after, she kindly introduced us to her husband, whom the Lounger found a very intelligent, though somewhat ponderous talker, with much information upon "the land" question, and decided views of his own, that the duty should never have been taken off from hops. And, later on, they give us a most hospitable invitation to visit them at Banbury Cross— which we feel compelled to decline for want of time, and regret the fact ever afterward.

As the night draws on outside, we sit by the window and look out on quiet St. Martin's le Grand, and on to the glittering corner of Cheapside, whose mufiled roar is ever in our ears. But oftener, just across the street, we watch the yard at the rear of the general post-office, and the red vans driving up by the dozen; each driver taking his accustomed place to get his special cargo of the nightmail, with which he will soon clatter off swiftly to his "own appointed station,"—one of the great railway stations of London, from whence depart the trains to every part of the kingdom, connecting with every corner of the earth.

And as you watch these, you will wonder which one of the vans is taking out your letter—the letter you wrote this afternoon to the dear folks at home, describing your first experiences in London. And then, just then, if you are a Lounger, you will have a feeling of home-sickness and loneliness come over you as a very small, strange unit in a very large babel and wilderness of a world. You almost wish that you could get on that van and follow your letter homeward yourself. "Warmest welcome," indeed! You know where that is to be found!

But the old couple from Banbury have already had their "jorum," their "night-cap" of brandy and hot water, brought in by the silent Will-yum, and have gone off to their rest. So you take your candle off the table, and stumble up another flight of stairs, and along narrow corridors with many turnings, with an occasional odd and unexpected step or two ascending or descending seemingly on their own account,--finding your own room at last, to go to sleep and dream of "home and friends around you."

* * * * * * * * *

This, you say is, however, but a Lounger's experience of several years agone, and things would be different there now! Not a bit of it. Things dont change down there in the heart of the Old City. "It may be for years, and it may be forever;"—but the next time the Lounger goes, he expects still to find the old "Queen's" just there, alongside of the "Angels," and at the mouth of "Bull and Mouth Street." And there will be the slender, prim old bar-maid, not looking a day older; and William, the Silent, will bring in the breakfast—the Times and the rack of toast, the fried sole, the water-cress and the beefsteak—in just that same slow, dreary way as of old! And the Lounger will ask him again, after lo! these many years: "William, what hour does the early morning train leave for Dover?" and William will answer: "I dont know, I'm sure sir, but there's the Times sir."

BEFORE DAWN IN OLD LONDON.

SAID to the Lounger on the eve of his departure for Europe, a friend who had been "over" several years before: "If you wish to hear some of the most extraordinary English you ever listened to in your life, be sure and go to Billingsgate Market and hear its fish-women abuse each other; but go at five o'clock in the morning, for that is their hour."

Just before leaving London for the Continent, the Lounger bethought himself of this advice, and anxious to prosecute his studies in Early English, laid out to lie down to rise up betimes, indeed, the next morning.

Springing from bed in the old Queen's Hotel, with the impression that he had overslept, he hastily consulted his watch thereon. "By the dawn's early light," which seemed breaking over the chimney-pots to the east, the hands indicated already a quarter past five! Hurriedly dressing, therefore, he stumbled down the steep stair-ways into the hall below. The dark statue of a man-at-arms in the front hall-way showed no surprise at this early advent, but the night-porter exhibited some in his countenance. The latter, however, cheerfully unchained the hall door, and the Lounger was soon speeding along St. Martin's le Grand, between the solid edifices of the General Post-office and the General Telegraph.

A few steps brought him in front of Sir Robert Peel's statue, on the spot formerly occupied by Cheapside

Cross, erected by Edward the First. A little to the right, in St. Paul's Church-yard, in ancient days stood "Powle's Cross," and just beyond its site, rose that vast pile, with its soaring dome which dominates Old London.

But the Lounger's course is first down the length of Cheapside, that thoroughfare which hitherto he had known so crowded and vociferous. Now, a wonderful transformation had come over that scene of multitudinous life and noisy activity. So silent, so deserted, one might almost be treading the streets of some Tadmor of the Desert, some excavated Pompeii of the buried Past! Had the Lounger chosen any other hour of the twenty-four, he could scarce have found such absolute quietude as now, when his solitary footfalls echoed on the silent pavement; and yet methinks some, yea, many, early goers might be abroad at half-past five of a summer's morning!

Impressed by this unwonted solitude and stillness, the Lounger pursued his way, between many a stately ware-room and home of daily traffic—past the sign of "Dombey & Son," ("tailors," it seems—alack for dignified Dombey, Senior, whom he had always imagined a great city merchant at wholesale) until between Friday street and Bread street, we recall that hereabout, on our right, was the site of the famous old Mermaid tavern, haunted by its Club founded by "rare Ben Jonson," and frequented also by Raleigh, Dr. Donne, Beaumont, Fletcher, and another playwright not wholly unknown to fame. Surely if ghosts may walk—if ever in streets of Old Rome those of earlier time came forth to "gibber and squeak" at night or by early cock-crow—this silent, deserted Cheapside should be the appropriate time and place for its numberless, historic, literary worthies to sally forth, once more to revisit "these traces of the moon," and of their own sadly obliterated footprints. Oh! if ghosts like these could only troop forth, this morn, "sheeted" but in the ink-stained quartos or folios of their own production, how gladly would the Lounger lay hold on Him of the Globe Theater, to stay his course until that fitful, fateful question should be decided, once for all: "Did you, or did you not write Hamlet, Lear, Macbeth, et cetera and so-forth?"

But the Lounger passed, and none of the Old Mermaid crew gave sign or token of their presence even in the spirit—in which, indeed, they were so apt to be at this hour of the morn when they had been "making a night of it."

Indeed the only presence that appeared to the Lounger was that of a stalwart British policeman who, shortly after, rose up before him, when he had strolled off up King street toward Guildhall, and was contemplating that edifice with the speculative intent of entering to look up those curious old wooden statutes of gigantic Gog and Magog. "Policeman X" looked rather suspiciously upon the stroller, as if he might be a prowling Fenian, intent upon blowing up, with dynamite, this lobe of the heart of London.

By the way, if the Lounger was after an anatomic figure, he might venture to divide the aforesaid Heart into the Auricles (not oracles) of Guildhall, and the Ventricles of the Mansion House,—with the civic life-currents of the Metropolis systematically pulsing between them.

Before this, however, the Lounger had left behind him, on the one hand, Bread street,—in which John Milton was born,—and, right opposite, Milk street, where Sir Thomas More first found experimentally what the street was named after,—and then passed St. Mary le Bow, whose bells once rang so plainly, "'Turn again, Whittington,'" but gave the Lounger no "turn" by admonishing him to go back! And there, close by, where the "big watch" sign projects over Cheapside, he got some solution of that problem of the silence; for the hands of that time-keeper showed the hour to be but 3:30,—and evidently he had inadvertently transposed, in his reading, the hands of his own watch, when he got out of bed at the Queen's!

What now to do? Why, keep on and see somewhat more of London in her night-cap—Old London before the early dawn! And here let the Lounger premise that, whilst nothing of adventure or of misadventure befell, the impression of the old town thus seen in its hour of preternatural stillness was itself a wonderful sensation, and one which the Lounger will not soon nor willingly forget. In the welcome absence of the hurrying, jostling thousands, he could roam and rove, hither and thither, or fast or slow, at his own sweet will, just as the roving lions roam about Persepolis—according to the Irish orator Phillips, and Briton Riviere.

Well then, he roamed on, past the Old Jewry and along where Cheapside narrows into the Poultry, that ancient "fowl" end of the old market street. And beyond the Poultry, as everybody knows, you come into that open hub of Old London, whence the spokes of streets radiate outward, and where stands the Mansion-House, the Royal Exchange (with the Great Duke's statute in front), and the Bank of England—that "hub" of the financial universe,—while nearly straight ahead is Lombard street, which is one of the biggest spokes in that wheel on which turn the world's exchanges.

But the Lounger turns down the fine thoroughfare of King William Street Instead. Just off this street, leads southward an insignificant alley, St. Swithin's Lane—whose mouth you might scarce discover, indeed—and a little way down this lane, by turning in under an archway, you come into a little open court, where facing you is Salter's Hall, one of the famous old Guild-Houses of London, while to the left is a long, low, plain-looking building that holds the most famous banking house of the world—Rothschild Brothers. The Lounger will scarce pause at this early hour to step into one of its little boxed-up reception rooms, where genial Mr. Silverthorn is wont to give him all his letters, as well as an instalment on his one Letter of Credit,—but, instead, keep on down King William Street, toward London Bridge.

Passing three or four other "lanes," when about half way down to the river, and at another open intersection—there where you behold the statue of smug King William IV.—once stood the old Boar's-Head Tavern, at the head of Eastcheap.

This, if we may believe the report of one William Shakespeare—and, truly, he was generally well versed in such—was once the resort of about as disreputable a crew as ever infested London. The landlady, a party by the name of Quickly—Mrs. Quickly, by courtesy—was, to say the least, no better than she should be nor half so well conducted as she should have been; while, for those she habitually entertained, we may say that, though some of them may have been high in rank, they were exceedingly "low down" in the way of morals—and of one in particular, "it were base flattery to call him a coward," a liar, a swindler, and a debauchee.

Pray heaven that none of their loose and ribald crowd

are abroad to-night—Sir John Falstaff and his swaggerers, the Wild Prince and Poins, with their crazy revellers—for what mad pranks they might play upon us! It is enough for a modest Lounger in literature to meet such wild ghosts even in play—of "Henry IV."

Now our street bends directly south toward the river. Through an opening to Fish Street Hill on our left, we see the tall Monument, erected to commemorate that blessing (in disguise) the Great Fire of London. At the next crossing below, London Bridge is just before us, flanked by flights of great stone steps descending to the Thames. Immediately to the left is the massive Hall of the "ancient and honorable" Fishmongers. Now the Lounger turns eastward, down Lower Thames Street. Chaucer, that past-master of Early English used to live here, you recollect, but has not now for some 500 years last past. And at last the Lounger arrives at Billingsgate Market; but lo, it is still dark and untenanted, and tall Policeman Y, who looks precisely like his brother, Policeman X, informs the Lounger that an hour later will be plenty early. So he strolls eastward still, past the Custom House, and on till he comes out into the open, squarely in front of the Lion's Gate of Her Majesty's Tower. At this hour he should scarce gain admittance, and so proceeds to circumnavigate it almost, passing on up the incline of Great Tower Hill.

If even in the broad light of day, the stranger finds the time-worn old fortress gloomy and stern, how might it now appear to the lone Lounger roaming about its long line of circumvallation, before the early dawn had imparted the first faint tinge of relieving glow! The impression now was one wholly in keeping with its history, and from Great Tower Hill, where its bloody scaffold stood, one could summon up some memories of iron hearts of the past, the noble or nobly born, who,— immured within its fearful prison, awaiting execution within its walls or the headsman's axe upon this hill,— saw life only as a fast fading vision, and death the great reality.

Sir William Wallace, the young Princes, King Henry VI., Duke Clarence, Sir Thomas More, Sir Thomas Cromwell, Anne Boleyn, Catherine Howard, Somerset, the Dudleys, Lady Jane Grey, Cranmer, Sir Thomas Wyatt, Southampton, Essex, Raleigh, Strafford, Sir William Russell; these are a few of the great names that serve in turn, to recall, not only their own tragic fate, but memorable epochs of that illustrious English History which is also ours by inheritance.

But by what irony of fate came it that William Penn, that apostle of peace and exemplifier of the blessed rule of good-will, was born on this terrible eminence of Tower Hill! Early his kind eyes looked down—as now the Lounger's in early morn—across moat, and over battlemented wall, and vast pile of wicked prison towers and inner fortifications,—with the White Tower, that massive Norman keep of William the Conqueror in the center;—and the lesson of all was that man should rule only by tyranny and oppression, through hate and through fear, with the aid of the prison and scaffold, by the might of the axe and the sword! The world is learning a far different lesson, that Penn helped to teach, but still it is learning it painfully slow.

Revolving such thoughts in his mind, and surveying the Tower from the different points of view, the Lounger passed around to the east into Little Tower Hill, and by the buildings of the Royal Mint, down past St. Katharine's

Docks, toward the Irongate and the Thames. A homeless vagrant lay stretched at full length on the stones of the walk, sleeping as soundly as might be. Policeman Z came along—or it might have been X or Y, they all looked so like twin brothers—and roused him up to "move on." Down near the Irongate, a waterman crept out of a hole in the wall where he had been reposing, and wanted to know if the Lounger would take a wherry. As he recognized the man at once for "Rogue Riderhood," he respectfully declined.

The Lounger now concluded he had traversed far enough in this direction, and so retraced his steps westward. It was past five o'clock when he regained Billingsgate, and the historic market was now open. He saw some cargo of fish landed at its wharf, and heard some fish auctioned off inside, but the whole thing was as commonplace as may be! The disappointment he experienced may be paralleled by the reader's sense of this feeblest of anti-climaxes, but he heard no wild, abusive language whatever, and in fact he saw no fish-women to speak of. In repayment for the loss of his morning nap, he had already secured some memorable impressions, but one more of the cherished illusions of youth had vanished,—fled, perhaps, to that fruitful limbo which preserves the apple of William Tell and the cherry-tree of George Washington. However, it may have been in the past, the fish-women of Billingsgate no longer "talk Billingsgate"—and there are no fish-women there, any way. They "may gang their ain gate," but it is no longer Billingsgate.

THE SALUTATION.

AMONG ladies traveling abroad these days, there has developed a habit or fashion (one might call it a *fad* if indulged in by the sterner sex) of collecting a silver spoon from each town that they visit. Appreciating this fact, the leading manufacturing silver-smiths of the continent were not slow to go into the business of supplying the demand, by providing a representative spoon for every town in Europe of any pretension to tourist travel, each spoon duly stamped, it may be, with the name of its sponsor city.

In due order of development, and following out the economic hint given by young Benjamin Franklin—who once irreverently advised his father to say grace over the whole barrel of pork when put down—the worldly-wise in their generation, who desire the biggest collection of town spoons at the least expenditure of time and money, would simply trace up this stream of silver to its source— the factories aforesaid—and there buy the whole outfit at wholesale figures. In this manner, the realistic bonanza-kings of silver are rapidly eliminating all the poetry and romance from life!

Now if the Lounger could have his way in making a souvenir-collection of European travel, he is not sure but he would prefer to bring home a choice selection of the inns he sojourned at when abroad. There would be two little difficulties in the way however. In the first place, inns hardly come within the category of what Dickens' Wemmick was prone to hanker after—"portable property." And again, the Lounger fears that as, after all, one chief charm of those hostelries that struck his fancy, was their agreeable harmony with their own quaint surroundings, they would lose much by being detached therefrom, and might "suffer a sea-change,"—into something neither so "rich nor strange."

Much the larger share of this fancied collection would be located on the soil of Old England. It would scarce include however, any of the larger or more modern caravanseries, though ever so complete or convenient and approximating the type of American hotels—such, for instance, as the Northwestern at Liverpool, St. Enoch's of Glasgow,—or, indeed, any of those moraines of railroad travel, the great Terminal Station hotels.

He would especially delight, on the other hand, in those that serve as reminders of the English Inn of the olden time. Outside of the old Queen's Hotel of London, one of the first in the collection, as it was early in the order of his experience, might be "The Salutation" at Ambleside. And yet possibly the Lounger is beguiled into this selection more from kindly memory of its good entertainment and some reflected glamour of its romantic surroundings, than from any other cause, for though two centuries old, the inn is beginning to take on extensions and modern airs, and to bid for tourist travel.

On landing at the head of Lake Windermere from the little toy steamer that had brought us up from Bowness, we had found the busses in waiting from the inns at Ambleside. It scarce took us long to decide in favor of the "Salutation." The homely title had a flavor of hearty welcome and of good cheer. How could it fail in its promise to the traveler, when its very name saluted him! And the assurance was fairly kept. The "Salutation" was a good house, homely and homelike, centrally located in Ambleside, which is a beautiful rambling village in the very center of the Lake District, which, in turn, is renowned as one of the most agreeably diversified and beautiful of all England. As the old toper insisted that there was no such thing as *poor* whisky, yet admitted that there were differences in the flavors,—so the Lounger would remark that England, while universally interesting to him, is not so entirely uniform in its characteristics of scenery as many people imagine. This corner of Westmoreland, looking over into Cumberland, is essentially Picturesque England. Beautiful lakes and waterfalls; green valleys and misty blue hills which at eventide draw near and project their wavy outlines against the sky, until you are almost ready to concede them the dignity, as they have all the poetry of mountains;—these are elements of charm that have drawn to them the feet of many who came with poetry in their hearts, and then felt constrained to abide here amid these scenes until that poetry found expression, which, in turn, has charmed the world. Wordsworth, Coleridge, Southey, Mrs. Hemans—as well as De Quincey, Kit North, Dr. Arnold and Harriet Martineau—these are noted names in English literature!

And this district has a nomenclature of its own for things picturesque such as are found within its borders. It provides its tourist seekers with "raises" in the way of regular ascents that can be overcome with roads,—with an abundance of "fells" and "scars" for rugged hills,— with "meres" and "waters" for peaceful lakes, and "tarns" for dark pools higher up among the hills,—with "ghylls" and "forces" for torrent and waterfall.

Quite in keeping with the local color of these names was the quaint appellation of the Salutation Hotel. We found our host quite genial and communicative for a native Englishman; one needs always to remember that qualification. He had been over to New York City, and had traveled on the Continent, so he was in a manner cosmopolitan. His rosy-cheeked daughter, who now kept the "books" inside the "bar," had received an education adequate to entertaining the tourist travel with which this district is thronged in the late summer, by being taken over to Switzerland and placed in a French school at Geneva.

Much entertaining gossip does he afford us concerning the district and its celebrities,—and then, following his direction, we are soon exploring some of the attractions of the locality for ourselves. A beautiful vale this about Ambleside, somewhat like the White Mountain Notch and the North Conway meadows combined in one picture! Over a stile, along a foot-path by the banks of the "beck," and then across the little river Rothay to the road beyond that circles past Fox-Howe, the old home of Dr. Arnold, still inhabited by his daughter, the sister of Matthew; and then skirting the foot of Loughridge Fell, the winding road brings us back at length into the village at its upper end, and past the entrance of Harriet Martineau's cottage, just off the highway and embowered in shrubbery—lilacs, laurels and pink rhododendrons now in all their blossoming glory of color. The two-story cottage is of the native dark stone, but covered to the top with ivy and sweet honey-suckle. We penetrate the flowery precincts of the enclosure, on assurance already given by mine host of the Salutation that it will be all right for Americans—and we are, in fact, most cordially received by Mr. Hills, who owns the cottage, and kindly shown all traces and memorials of herself left behind by Miss Martineau, in this secluded retreat, where, amid the beauty and bloom of nature, the woman of great intellect patiently awaited the end of so many years of suffering. In the library was still the little bust of Clytie over the center of the mahogany shelves that had held her books, and the rug on the floor given by Jacob Bright, while outside, below the circling terrace, where the slope fell away toward the "beck," stood her sun-dial with its pathetic inscription, "Light! Come visit me!"

And when after another stroll, up the hills to Stock-Ghyll Force, we return rather wearied, the good hostess of the Salutation meets us in the doorway, with her own kindly salutation, and—What would we be pleased to have for supper? In the absence of any expressed preference, she herself finally suggests—would we like a nice char? Now this was another new word in the Westmoreland vocabulary, concealing we knew not what in the gastronomic repertory—but the Lounger fell into hearty acquiescence:—Yes, that would be the very thing! For aught he knew, it might belong to either roast or boiled,—in the category of patties or of puddings. A dark suspicion even crossed his mind that it might prove one of the protean names or forms of hotel hash! But we were in for it, all the same, and with a tranquil mind awaited developments.

The sequence was a happy one. The char was not charred, but done to a turn, and proved to be a serving of the most delicious and delicate a broil of fish that ever tired travelers did justice to at close of day. Justice, say we? Yea, for more than justice, for the bones were picked clean!

To the Lounger's inquiry if these were not trout indeed, the landlady insisted to the contrary. It was "Char"— a fish caught here in Lake Windermere, on this last Mayday of the season. "A month hence and a guinea a piece would not secure one," she averred; as also that trout could not live in the same stream with char. She was not so far wrong, for if the "proper authorities" are to be relied on, whilst char are of the trout family, they belong not at all to the true genus "trout"—any more than do our American brook trout which are actually a species of "char."

However, instead of raising any of these nice questions in Natural History at the time, we preferred to petition the hostess to raise us some more char for the morrow's breakfast;—and again we feasted royally upon them for first course, perhaps "not wisely but too well," indeed, for when the waiter followed after with an ample supply of lamb chops, tender and succulent, we fairly confessed our inability to do that subject justice. The mildly reproachful look of that waiter—somewhat tempered, perhaps, with pity and contempt for our incapacity— lingers yet upon the Lounger's memory after lo, these many years!

And we left at length and most regretfully the Salutation and the shores of beautiful Lake Windermere. We chose not the route by Kirkstone Pass, which leads by what is sometimes pronounced the "'Ighest hin-'abited 'ouse hin hall Hingland,'—but the road across "Dunnail Raise" and on to Keswick. A delightful ride this, which led us first between the hills, and past Rydal Mere (a *mere* pond no larger than a mill-dam) on the one side, and Rydal Mount, Wordsworth's old place, opposite; then by Grassmere Lake and Grassmere Village, stopping awhile at its little church-yard to note the graves of the Poet and of his daughter Dora; thence on again, up the long ascent of the "Raise," and skirting the western flanks of the "Mighty Helvellyn;" then for a few miles by narrow Thirlmere, whose pure waters have been bought up, to be conducted for a hundred miles across the country to thirsty Manchester. Then coursing along valleys between the ranges, we finally came to where Saddleback and Skiddaw rose up in front, and from the summit of a slope could look right down upon Derwentwater with other beautiful lakes shining among the hills; the pleasant little town of Keswick, where Southey used to dwell, nestling in its nook, with picturesque, misty mountains standing around to "sentinel enchanted ground." And there, on the shores of blue Derwentwater, near the Falls of Lodore, where the waters "come down" with all their many particles and participles (ending in *ing*)—when there are any waters to descend—we find another inn, picturesquely situated, indeed, but by no means sufficient in charm to displace or dispel the memory of Ambleside and of The Salutation.

MORE OLD INNS.

BEFORE resuming in good earnest the cataloguing of his souvenir collection of English Inns, the Lounger has a mind to advert for a moment to some specimens of a different type. For these, it is true, he owns not the same measure of affection—indeed he doubts whether he loves them at all—but they, too, are historic in their way, and interesting in their survivals. These British hostelries of a generation or two agone mark the transition period between the old English stage-coach inn of the past, and the commercial-traveler and tourist emporium of to-day. While the first, in its day, was prone to heartiness and jollity, and the last gravitates inevitably toward the noisy bustle and activity of the American caravansery, the intervening type was sedate and eminently respectable always, and sometimes almost preternaturally solemn.

Of this character, the Regent's Hotel at Leamington was a conspicuous example. How intensely dignified and how ineffably dreary it was—and how fearfully ponderous and respectably stupid were its chuffy waiters, in their swallow-tailed coats and high cravats. The one redeeming feature of the hotel was its exemplary dining-room, with its side-board so massive, yet fairly groaning under the weight of those noble joints borne in by the fat, short-legged waiters, bending under the burden of their immense platters as did Atlas with the World upon his shoulders.

Then too, for another specimen,—the County Hotel at Carlisle, just outside the old walls and across from the station! This is a fair example of that high-toned institution as depicted in the modern English novel,—the place where the advocates assemble during the Sessions, and where the gay young belles of the gentry sometimes pass the remaining hours of the night, after their attendance at the annual County Ball. Its rooms are hung with old-fashioned engravings, and its corridors with oil-portraits, in massive gilt frames, of exalted personages of the Royal Family for many generations. Most excellent—and stupid—is the County Hotel of Carlisle!

But when we want a genuine old country inn of the primitive type, commend us to "the Lowdoun," at Mauchline, in Scotland. Than this little village probably no better representative of the Lowland Scotch country town could be found anywhere, and its sole inn is in thorough keeping with it. We had stopped off here to see the village that Robert Burns had frequented, the one whose scenes and personages he had commemorated, and from which, on the publication of his first volume of poems, he suddenly burst forth upon Scotland and the world as the meteor genius of his time.

Apart however from its association with Burns, we were glad that we had visited Mauchline, feeling that we had caught a little glimpse of the old-time rural life of Scotland, almost fossilized in this little out-of-the-way village. Especially characteristic of those times was this rambling old inn, with its thatch roof mossy with age, and delightfully picturesque in the afternoon sunlight. Its interior too, might afford fine scope for the painter. We penetrated the old kitchen, with its stone floor and generous fire-place, and found its utensils and the furniture generally characteristically antique. In the tap-room, which we traverse on our way to the lunch-room, a pair of rustics are discussing a tankard of home-brewed, in regular "genre" fashion. Here, after visiting the old Burns' farm of Mossgiel, we did full justice to the homely lunch, including cuts from a noble round of cold roast beef, pickles, oaten cakes as thin as wafers, cheese and rhubarb tart. And then we go on our way rejoicingly to "Auld Ayr," Kirk Alloway and Bonnie Doon.

* * * * * * * * *

Once,—in the good old days "when George the Third was king"—a highly fashionable watering-place, and still, though deserted by the throng that crowds to newer resorts, a haunt of British dowagers of the most eminent respectability, Leamington Spa aforementioned boasts a location wholly unsurpassed for elements of pastoral beauty, for it is in the center of Warwickshire, the geographic center of the country, and the very heart of "Merrie England."

And yet, lying thus in the midst of the loveliest and most cultivated country in the world, affluent with glade and stream and bosky dell,—where the foliage is of the densest, and the verdure of the brightest emerald; surrounded by parks and aristocratic country-seats, with attractive towns and villages on every hand,—Leamington is scarce sought by the American traveler but for its railroad station and its convenient access to its more famous neighbors, romantic Kenilworth, historic Warwick and word-renowned Stratford-upon-Avon.

As the Lounger is not here a chronicler of the glories of the eventful past, he will scarce indulge in a retrospect of his visit to these memorable scenes—and after the grand round past the historic castles, will pause but for a moment at the pastry-cook's shop in the stony High Street of Warwick village, and then travel on by green lanes and hawthorn hedges of blossoming white and rarer pink, to the Red Horse Inn at Stratford.

The halt at the pasry-cook's was, of course, to bait ourselves with a glass of milk and a bun. While indulging in this refreshment, our driver came rushing in, breathless with the excitement of momentous news: "His Ludship, the Hearl of Warrick is comin' down the street!" Willing to sample the quality of that "divinity which doth hedge in" a live Earl upon his native heath, we moved to the open doorway—and there, came ambling down the sidewalk with tottering, teetering steps, a feeble, superannuated old London beau! We say to ourselves, "Oh! how unlike the historic Kingmaker of the olden-time whose title you inherit. Though the descendant "of a hundred earls" (which is more than doubtful) and with all the treasures of yonder stately pile, at your back—"You are not one to be desired!"

After Shakespeare and Washington Irving, it were surely idle for the Lounger to attempt to paint the lily of the Avon meadows, or to gild the refined gold-leaf of the old sign of the Red Horse Inn. It may suffice to say of the latter, that the charm of quaintness which Irving found therein is still measurably preserved in the irregular, rambling old tavern, together with the memory of Irving himself, whose literary commendation has been an interest-bearing capital ever since to the Inn. They do well, therefore, to call the little seven-by-nine room on the ground-floor, just off "the bar," the "Irving Parlor;" with his arm-chair and a few other relics of the author's stay faithfully preserved; and to seat the Lounger therein for supper, after he had visited the old house in

Henley Street, where Shakespeare was born, the old Guildhall Grammar School, the site of New Place--and had strolled, over stile and by foot-path, across the little fields to Anne Hathaway's cottage at Shottery. Shakespeare went for Anne across lots you remember.

Yes, we had done all this in the late hours of one September afternoon; beside visiting the Old Church where he lies buried under that slab whose doggerel inscription is certainly a grammatic oddity---whether crypt-o-grammatic or no.

And the next day, which was Sunday, we spent as a day of rest at Stratford. When the church bells were ringing for service in that same old church, the Lounger fears that, true to his name, he was rambling adown the meadows, past the old mill and along the banks of the peaceful Avon, which for all the world, so far as natural characteristics go, might have been the beautiful Brandywine instead. And the Lounger wandered and pondered— "loafed and invited his soul"—and then came back and had a homely dinner at the Red Horse Inn. Did he muse much upon Shakespeare, and gain any new light upon the enigma of his being and doing, here amid his early and later haunts—where the rude cottage of his birth-place still survives, whilst the fine house which he built on the best corner-lot in town has long since vanished? Well, the Lounger fears that all the light he gained here was but darkness—made visible!

And lo! as the Lounger mused thus at eventide, on the High Street of old Stratford Village, in front of its old Grammar School, where Shakespeare learned a great deal of Latin--or learned none whatever, (the world is not quite sure which)—a modern Minstrel of Song came along, with his wife and children three—minor minstrels;

all leading each other by the hand, strung across the street as they passed by. And they lifted up their voices and sang. It was none of the melodious sonnets of Shakespeare; no catch from madrigal or roundelay of the Bard of Avon—but the first line thereof had a familiar sound notwithstanding: "There's a land that is fairer than day!" And the Lounger, who had then been many long months away from home, said to himself: "Yes, I know that land right well—it's America"—and went and dropped a nickel (or was it four-pence) in the hat.

* * * * * * * * *

Only one more old English inn, and then the reader shall have his "innings," for a spell! 'Tis the Peacock Inn at Rowsley. We are in Derbyshire now, but Merrie England stretches over this way and includes this region, beyond question. From Rowsley you can have your choice between the choicest of the Old and the New of England. If you are of the mind of the Lounger you will scarce fail to choose both. Two miles up the Wye, and you are at Haddon Hall,—perhaps the best type, and at least the best preserved, of the Old Baronial Hall of England. It has not been inhabited for 150 years, but here it is, all in apple-pie order, kept thus for you and me as an object-lesson happily illustrating Old English life. Here is the old kitchen, with the slaughter-room attached. There they killed the ox, and in the adjoining kitchen, on the spits before the immense fire-place, they roasted him whole, when "the Baron and all his retainers gay were holding a Christmas holiday."

And next is the old Banquetting Hall (of the thirteenth-century) with its stone floor, its massive slab table, worn and time-eaten, with its rude benches of slabs for seats; and, though the hall is low-storied, yet, stretched across above, is the gallery for the minstrels. The ox (almost as large as an elephant) "now goes round—the band begins to play." Farther along is the dining-hall of later date—fifteenth century now—with its cross-beam ceiling, hacked in after time by Cromwell's soldiers. And many other rooms there are that belonged to the two differing eras of its building. It is all wonderfully interesting "from turret to foundation stone," even to the postern-door, leading out on the terrace, through which Dorothy Vernon eloped, and carried the estate into the family of the Duke of Rutland, which holds it to this day. These, and a hundred other things, they show and tell you—all wonderfully cheap for your shilling to the house-keeper's fair daughter.

But on the other hand, only four miles from Rowsley, along the Derwent, is Chatsworth, the great seat of the Duke of Devonshire, and one of the finest show places in England—or the world, for that matter. If Haddon Hall possesses all the interest of antiquity, Chatsworth boasts the glories of modern days. It is full of magnificence—frescoes, carvings, royal furniture and bric-a-brac, paintings and statuary;—and all by world-renowned masters in each department of art. Then, too, the finest conservatories, gardens and park, everything on a grand scale. They show you an India-rubber tree, which discharges a shower of water when they get you planted —under it. A veritable hoax, you see! They point out also oak trees, said to have been planted by the Queen, and the Czar of Russia. These, may be "Royal H-oaks" also, for aught the Lounger knows.

But after all is said and "done," you will enjoy your night's rest succeeding, at the old Peacock Inn. A most picturesque old inn, indeed, with its gables, its latticed windows with diamond panes, its clinging ivy outside, and its cheering fire-place indoors. All these things are built to order now-a-days; but this is the genuine old-time thing. And then its pleasant little chamber overhead, where they put you to rest amid such abundance of clean and fragrant linen. It really seemed that no inn in Britain was too small to own a profuse supply of fresh, wholesome linen! You have read Leigh Hunt's essay on "Bed?". Well, it has always seemed to the Lounger that Hunt's essay could scarce have been written in any country in Europe save England.

And over the door of the old Peacock Inn was a venerable inscription. The Lounger took it at first for some cryptogrammic Old English of Robin Hood and Friar Tuck significance, that might somehow mean "Honest Venison." But he will just leave it to the reader to decipher. Here it is:

IOHNSTE
16   53
VENSON

FRENCH AS SHE IS SPOKE.

THE LOUNGER had just returned to Paris from the Rhine country and, strolling along the Boulevards des Italiens, the next morning, had stopped for a moment at the end of one of those *Passages* that lead into that thoroughfare, partly to take shelter from a passing shower, but more to use that pretext in order to observe interestedly the flow of the life-tide along the avenue,—not swift and tumultuous as in Cheapside or the Strand of London,— but rolling with the gentle swell of summer seas. Your genuine strolling Parisian takes life as if it was all before him, one long holiday—except indeed, when perforce darting swiftly for a moment, to make the dangerous transit of the crossings and escape the present peril of the reckless *fiacre* driver.

Among the approaching throng, however, the Lounger's eye was attracted by a conspicuous exception to the general rule of the current,—a green bough of driftwood amid the eddying stream. Not only in the elbowing, nervous action and swinging stride, hurrying to get there— not alone in the restless, roving eye and the sharpened physiognomy—but in a dozen other ways, indefinable but patent to the senses, the Lounger recognized at once a familiar type, and mentally ejaculated; "My Country, 'tis of thee!"

This recognition needed not to be supplemented by the customary chronic-catarrhal tone of voice, but this was supplied, for the stranger, seeing the Lounger at ease, deemed him just the person nearest at hand to ply with the question nearest to mind—so promptly brought up before him with: *"Mossoo, vooly voo, me deary-jay ah Drexels, sivoo play!"* The Lounger determined not to 'play'—at talking French—so replied with his usual formula: *"Je ne parle pas le Francais."*

Something in the Lounger's accents, in spite of their refined tone, seemed to betray a kindred nationality to the stranger, who, scanning his face sharply, quickly responded: "Well, you talk American, dont you?"

"Like a native," replied the Lounger.

"Well—I'm mighty glad of it—where's Drexel's? But never mind—that can wait. What State are you from? How long have you been over? What boat did you come on? Kansas! Why, you dont say! How lucky I spoke to you! Why, I'm a Sunflower myself—you bet I am. Brattle's my name—of Ingleson. What's your town? Lawrence! You dont say! Why, I know the 'Old Historic' like a book,—especially some of the University folks. I'm a teacher myself—educated at Ballburn College—and I've been runnin' the Fourth Ward School at Ingleson. Came out on the Anchor Line to Glasgow. Been over six weeks already. Our folks at Ingleson are goin' to give me First Assistant in the High-School this fall. Just come over to rub up my talkin' French a little, first—I'm to teach it, you know. Had a thorough course in it at Ballburn and read it, like a book—Telemack and all the rest—and now I've got on to the pronunciation first-rate. I can parley-voo with any of 'em. Of course the genders bother me some, there's so many things engendered here in France, But what did you say your name was? Lounger! Well,—

there's a large family of you in Lawrence, isn't there! Ingleson's the town, though;—pretty much all the noted men of the state come from Ingleson."

And so he ran on, apparently delighted. "But say, let's go somewhere where we can talk. In one of the Caffys, say? Why I hadn't seen the first blessed Jayhawk since I left home. Come, let's have somethin' together, right away."

"Much obliged to you, Mr. Brattle—I'm glad to meet you and will go with you gladly—but bear in mind, that collectively you and I represent a prohibition State, and as personally you are a teacher, it would scarce be proper for us to be seen drinking in cafés."

"Oh bother that," exclaimed Mr. Brattle with some freedom; "we're a long ways from home—and you know what the apostle enjoined—'when you are in Rome do as the Romans, and be all things to all men.' Well I'm a roamin' now—and a little good wine won't hurt us."

"That may all be, Mr. Brattle, but I never drink in cafés. One is compelled to draw the line somewhere— and I draw it around the dinner-table exclusively. In the absence of a certified analysis from the chemists of the State University as to the total freedom from microbes of the Seine water, I do feel compelled occasionally to take, at dinner, one small"—

—"*Wee, wee*" broke in Mr. Brattle "*Jer cumprenny— oon demmy bootell de vang ordenare.*"

"By no means, Mr. Brattle," quoth the Lounger with some dignity; "always 'the Chambertin with yellow seal,' as Thackeray says."

"Well—come along anyway. There's plenty of Caffys around here—I know 'em pretty well. There's one thing that isn't disappointin' about Parree, and that's the

Caffys. I've got the *ong-tray* of them all, and I find 'em *trays interessong*, especially those in the Shams-e-lizzy. Which will you have now—Caffy Reesh, Tortony's or the May-Song Dory? Well, let's try Tortony's, then; it's near by and first-class."

"Here we are"—he ran on as we moved on—"let's go inside; no sittin' out on the sidewalk for me. It looks too public and second-classy for an American. Besides, it's too near the curb and those things. They may be all right in one sense, but they're all wrong in other scents. Queer people these French, dont you think? Just look at 'em drinking their green absynth! Wormwood and gall,—I've tried it and I know. It'll kill 'em off, sure. Here we are inside and here's a table! Hope you did'nt forget to bow to the *dam de comptwor* there at her desk! Sounds like a title of nobility dont it? What, positively no *vang!*—then what'll you have—here's the maynoo."

The Lounger referred it back to himself.—"Well, here's '*Fromazh glassy;*' fromazh means cheese, Lounger. Sorry you dont talk French, its easy enough. Dont see how you possibly get along without. Do you like cheese? I'm mighty fond of it, myself. Well, let's have some of this, and a roll to go with it. '*Garsong!*'—'garsong' means 'boy,' but you always call the waiter 'garsong' even if he is baldheaded—'*garsong, apportong noo du fromazh glassy, ay du pang, poor doo! eh beang!—attonday, garsong!*' '*Sorebay de crame ah lah Vaneel*'—that must be somethin' nice for dessert. *Sorebay poor doo, garsong—partong! allong!*"

The Lounger now thought it his turn: "Where are you stopping, Mr. Brattle?"

"Well, I did try the 'Hotel Days Aytah Unee' in the Roo Show-Say Darntarng; the home name took me in, you see. But I've changed now to the 'Trwaaw Prangse' in the 'Petty Shams.' It's a good house, and though the name's aristocratic (it means 'Three Princes,' you know) the prices aint. You should have seen me put my foot down on the *port-yay* and the *sairvongs*, the whole kit and bilen' of 'em, when I came away from that first hotel! They got no fee—no *poor-bwor* from me— on principle! I just said to the port-yays 'you drop that valise, I'm attendin' to that myself,'—said it in good American—and they let me alone. But most always I get my dinners out, at the '*Aytableeshmong Bullion*' in *Roo Cat-Set-Tom. Deenay ah pre-fix* you know."

"But inasmuch as you are so fond of Caffy, I rather wonder you do not patronize the '*Hotel de Veal*,'" quoth the Lounger.

"There it is now, Lounger; that mistake of yours just comes from your not knowin' French. That name simply means the same as 'City Hotel' in our country. Then there's the 'hotel days Arnvaleed'—'Invalids,' we would call it; just a kind of family hotel, you know, for those who want to rest up and recuperate their health."

"Of course you've seen many of the sights of Paris already, Mr. Brattle?"

"Seen the most of 'em, I guess, Mr. Lounger. Didn't come here to let the grass grow under my feet."

"Then you have 'done' the picture galleries and seen the famous pictures and statues?"

"Statues by the ton and pictures by the mile—in the Loover. Saw that, one of the first things. I was rather 'fresh,' then, and hadn't got the lingo down just so fine as I have now. I had quite a hard time, indeed, gettin' into the Loover. As I was directed, I started from the Bullyvars here, down St. Lewee le Grong, across Roo

Cat-Set-Tom, into Avenoo Lopera. That took me down, you see, into Plass Tayater Frongsay. There I stuck; so, I spoke to a boy—one of the Parree 'game-uns,' he was—and tried to get him for a guide. I said: '*Garsong, mong bong ongfong, mongtray mwaw le Loover, sivoo play!*' But he only shook his head and held out his hand, till I put a frank in it, and then he ran away. Then I asked a John Darm, and he took me, for 'twas only a step, into the Roo de Rivoly, and showed me the doorway. Well, when I entered the old palace I found I was in the wrong box somehow. There was a big lot of clerks and desks, but not a single picture nor statue. If you'll believe me, Lounger, 'twas the Departmong of Feenongse of France! I stepped up to a head-man there—he may have been the Minister himself—and I said: '*Mossoo, esker say le Loover easy, sivoo play?*' and he said '*wee.*' Then I said '*May—oo ay lay paint-your ay lay stat-your,*' for I had forgotten the right French for statues. Then he smiled, and talked a blue streak for a minute, without any punctuation-marks between the words or sentences; and he made me tired, for I couldn't get the hang of what he was sayin' worth a cent, only he kept repeatin' somethin' about '*Mu-say.*' When he stopped at last, I said '*Jenny come prong pa:—Jer swee Americain. Jay oon grong day-seer de vwor lay Loover— lay paint-your ay lay stat-your*'—and as he smiled so pleasantly, I thought I would give him some 'taffy' in my best French—so, I added, '*Mossoo, jer swee sharmay de voo vwor!*' Then he bowed and smiled again, took me by the arm, led me to the opposite door, opening on an inner court, and pointed across, to the entrance of the Musay. Oh, but he was a polite man for a Minister of Finances—if he was the minister."

Just then the waiter brought in the order—the 'fromazh' and the 'sorebays'—and, to the astonishment of Mr. Brattle, his loved cheese was absent, while outside of the *pang*, the dishes seemed to be almost a duplication, and of some half-frozen substance very much resembling the ice-cream of our own confectioners. Solacing ourselves, however, for our disappointment, by the reflection that we had, at all events, secured our national luxury, we proceeded to enjoy the refreshment—and, Brattle, with his reminiscence. "Well, I saw the statuary in the basement, the Venus de Meelo and all the rest, most of 'em badly mutilated and demoralized, and the thousands of pictures above, especially the renowned ones in the Saloon Carry. But I didn't think much of the buildin' itself, Lounger; it's a dingy old thing, with a mansard roof, such as we tried in our country years ago and discarded."

"I'm sorry you didn't like the architecture," quoth the Lounger, "though it is of rather a gloomy character. But have you yet seen the Madeleine—the beautiful Madeleine?"

"Well, I hardly know, Lounger, but I guess I have. I went to the Caffy Shon-tongs and the Caffy Dong-songs, in the Shams-e-lizzy, the night of my arrival—and I think that was the name of one of them. I tell you 'twas high-jinks there Lounger;" and he winked audibly.

"Oh Brattle," exclaimed the Lounger, "you have seen far too much already in this wicked city of Paris, but the Madeleine is not its Mabille, but one of its sacred things instead! It is a cathedral, one of the most beautiful of churches."

"Well, then, I hav'nt seen it, and I do recollect now it was Mabel the other affair was called, the Schardang

Mabel; but speakin' of churches, I tell you, Lounger, I have seen Not-er Dam Cathedral, in the city."

"Brattle," quoth the Lounger severely, "I fear that evil communication has got in its fell work already; but no profanity, please! Remember that at home you are an instructor of youth, both on week-days and on Sunday, and there should be a limit of latitude, when abroad, even for church members. They shouldn't travel too fast nor too far!"

"Why, Lounger, I wasn't swearin' by no means, and if I was off about the Madeleine, you are now about Not-er Dam. Why that's a sure-enough church too, and they told me it was 'dong la City,' though nobody ever supposed it was a meetin'-house out in the country, I reckon."

"Oh, they meant it was in the Old City of Paris, Mr. Brattle,—across the Seine—Ile de la Cité."

"But, Lounger, I want to know how you caught on so quick to my bein' a Sabbath-School teacher back at home in Ingleson?"

"Why you gave yourself away, Mr. Brattle, when you acknowledged that you went to the Mabille the very night of your arrival. You extra good people are so anxious to see how wicked a city is Paris, that, as you say, you let no grass grow under your feet. The rest of us can afford to wait awhile, and then, perhaps, forget it entirely."

"Well, that's all right, but you've got to see for yourself; and, one thing I can tell you, that a good deal of that High Art in the Loover—'high old art,' I call it—such as those pictures of Ruben's in the long gallery discounts the Mabel people badly, while the Shon-tongs folks are modest in comparison."

"Your criticism is just, Mr. Brattle, and modern French art is no improvement. Possibly you are not so well versed in that as you are in the language! It must be a great satisfaction to speak that so fluently."

"You're just right there, Lounger. It's a great advantage, especially in the shops. They just let me have things away down. The shop-girls say my accent is *'parfate, o mar-vale.'* I see 'em smilin' at each other sometimes, it pleases 'em so to hear their tongue spoken so well by a foreigner. Some of 'em take me for a Parisian born. Oh, they're just as bright as Americans! I came across a stupid one the other day, though, in a shop in the Roo de lah Pay, where I'd gone to buy a fan for my *feearn-say*, back in Ingleson. The girl couldn't understand me, any more'n I could get on to her French. After jabberin' at each other for awhile, she suddenly broke out in good English, and then it came out that she was an English-woman, from London. Hadn't been over more'n six months. But, I tell you, Lounger, you just want me to help you buy things here. I'll do it."

By this time our refreshments were exhausted, and, giving Mr. Brattle my hotel address—and the direction to Drexel's—we reluctantly took leave of each other for the time being. " *Wee, wee,* I'll look you up *toot-sweet* at the St. James," said Mr. Brattle, "and we'll have some good times together. I'll show you around; but it's rather a pity you dont know French, Lounger. *O ravvwor, ardeoo ah pray-song!*"

CONTINENTAL HOTELS.

LONG years ago the Lounger made and heralded the notable discovery that the inns were just one-half "the ins and outs" of travel. Did this even apply solely to the "money out" it were an announcement worthy, perhaps, to go down to remote posterity parallel with that other discovery, of a great economic genius, that "the butter of a family costs more than its bread."

The reader will, perhaps, suspect that the Lounger is inclined to emphasize the importance of this subject of inns, as some manner of apology for devoting so much attention to those of the Old World, in particular. This paper will be devoted to inns in general—or rather to some general characteristics of the Continental Hotel.

As his omnibus or carriage from the railway-station rolls into the hotel court-yard, or sweeps up, with a cracking of whips and a final flourish, before the hotel entrance, the neophyte in European travel is apt to be a little oppressed as well as impressed by the grand style of his advent. This in spite of the fact that Matthew Arnold always advised his readers—and writers—to cultivate "the grand style." Oppressed, indeed, in this instance, not only on account of intrinsic modesty, but with a vague suspicion that some of the grandeur and warmth of this reception may come to figure in the bill. He has heard, mayhap, that honors are not so easy in Europe, and that everything has to be paid for, sooner or later. Here are assembled apparently a majority of all the attachés of the establishment, vying with one another in their eager attentions to the traveler, and seeming to say to him, as in the mute but expressive language of the wagging tails of the house-dogs at the parsonage of Praed's Vicar—"Our master knows you—you're expected." Smart, discerning people, these Continental landlords—but how is it possible that they got hold of the fact that a member of the Town Council of Smithville was coming on this train, when you hadn't even telegraphed ahead for rooms?

On entering the doorway, here is the greeting landlord himself, with the rest of his subordinates—if, indeed, it has not chanced to be your fortune, as sometimes that of the Lounger, to find them drawn up in parallel lines on either hand outside. Conspicuous amid the throng is that intelligent and distinguished looking gentleman with a gilt band on his cap, whom you soon come to recognize as the factotum of the establishment, worth all the rest put together, and ready to be ranked your "guide, philosopher and friend," the *portier*.

Do not get excited at all this honorary array, however; it is not peculiar to yourself alone, and for yourself, it will never greet you in such force again but once—when you come to leave! It is simply a hotel contrivance to "welcome the coming, speed the parting guest."

Thank fortune, however, there is one individual (the Lounger might more haply term him 'institution') that is lacking,—that haughty and much-be-diamonded-in-shirt-studs tyrant, the traveler's born lord and master, the American hotel clerk. He, who, like the Speaker of the House of Representatives at Washington, recognizes solely whom he pleases; and again like said Speaker, cares to count you in only when he wants to make a quorum;—he whom you vainly implore, for a quarter of an hour, to recognize you, on the floor before him, while your poor, tired, head-achy wife sits forlorn in her weariness and traveling-duster, up in the reception-room; —he, who, when at last vouchsafing to behold you, studies his key-board for ten mortal minutes longer, in order to pick out and assign you the worst room in the building, away up under the eaves, while he retains the best in the house for some commercial-traveler crony of his own, who is coming along by and by.

Instead of this, up steps, the moment you enter, a major-domo, with his list of vacant rooms in his hands, at his finger-ends,—and in two minutes your rooms are assigned you, and within three more, you are installed therein. *Then*, the waiter takes down your name and number, which are entered on the *portier's* book in the hall. They rather like to have your name, as it is convenient in case of letters coming, or acquaintances calling for you; but it is not indispensible otherwise, and you may, if you choose, be known only as "37," or "the lady and gentleman in 68." You discover, after a while, that they really know—or care—nothing about that Smithville Alderman business after all. It will scarce be worth while to impart to them that you expect, indeed, to be chosen Mayor next year. A mark or a franc piece will go further with any waiter in Europe than any such piece of frankness or mark of confidence on your part. You need not assume any factitious dignity, however. Keep your reserves for the landlord—but treat the waiter frankly!

Once at Cologne, the Lounger was quartered in a room adjoining that occupied by one of our leading Democratic candidates for the presidency, who received no more honors or attentions there than did the Lounger himself. In fact, none of the hotel-folk had ever heard of him before,—not even the "portier."

Well, you get your room, and soon you ring the bell. If you do that the first thing on your arrival in an American hotel, the evening of a hot day, unmindful of the bell-code and the servant's convenience—the waiter who toils up those weary flights of stairs, takes chances on knowing what you want by intuition—and brings a pitcher of ice-water for drinking. Your English waiter thinks he knows what an English guest would wish under such circumstances and fetches you a jug of hot water for bathing. "Such is the custom of Branksome Hall!" Your Continental waiter comes bearing a fresh pair of candles, and lights them—which will lighten your purse to the extent of two-francs for every evening. The landlord has no notion of allowing you to follow in the footsteps of the sire of "Proud Miss MacBride," who—

"Little by little grew to be rich,
By saving of candle ends, and sich."

Taking it all in all, the European hostelry compares not unfavorably with our own. The larger and newer are usually supplied with all modern conveniences, except toilet-soap, and this, like the candles, you can have for 'a consideration.' As a rule, their hotels are not nearly so spacious as our largest caravanseries, and less profuse in extravagant decoration. In large cities, like London, Paris or Berlin, one can find some of the same description. The Grand Hotel, of Paris, and the Central, of Berlin, occupy each, quite a "block," and their cost ran into the millions, as well as that of the ground they occupy. Moderate-sized hotels, however, are the rule. In these, the space devoted to public rooms,—parlors, reading-rooms, etc.,—seems quite abridged to our extravagant notions. Such suites of public parlors as we find in our grandest hotels are rare indeed. In the Lounger's hotel off the Rue St. Honoré, in Paris, reception-room, parlor and reading-room were all one, and frequented by guests of both sexes. It was a comfortable hotel, largely patronized by English families. The lack of these fine rooms is often largely compensated by the inner-court or inclosed square, which is beautified with turf and trees and flowers. This inner-court is beginning to be a feature of recently built American hotels also. The rooms opening on this are far more desirable than those on the noisy streets, abroad. The dining-room, opening on such a flower-garden, with sometimes a fountain in play, is especially pleasant. The Lounger found even in some of the dingiest and dirtiest continental towns, about the hotels and elsewhere, more evidence of a taste for flowers than is usually exhibited in our esthetic country. Scarce any hotel dining-table was considered *au fait* without them, and often the staircases were adorned with rare plants and blooming exotics.

Whilst at first the untraveled American hardly takes kindly to the European system of hotel charges, he often comes to regard it as, after all, the fairest and best. He pays for his room according to its location, and with his meals, for what he orders,—at least in theory. In practice, unless he keeps some track of his orders and the items of his bill as rendered, he may pay for something more. But there is a tariff of fixed charges for almost everything you are likely to want, and it will afford you pleasant exercise to commit it to memory! Then, "cut your coat according to your cloth,"—being sure that there will be plenty of those who "stand and wait" around for the scraps and pieces left over. Try and match the head-waiter's book-keeping with your knowledge of arithmetic—and you may come out nearly even!

One of the "institutions" on the tourist line of travel abroad is the "table d'hôte" dinner—a fixed-course meal at a fixed hour, at a fixed price. Your American abroad for the first time is sure to take it. If there is any table in Europe that he dotes on it is the table d'hôte, for it reminds him somewhat of the public hotel table at home. Your reserved Englishman may sit down at table aside, to his special dinner ordered from a few favorite and well-cooked dishes, preferring such to all the long rigmarole of stewed, roast and boiled. The Lounger deems that he has the best of it. He is a little "out of touch" thereby with the throng of travelers, but what he possibly loses in sociability, he gains in digestibility. With slight modifications, the same "table d'hôte" dinner pervades tourist-Europe. The Lounger found it (to his sorrow) a conventional dinner, invented, perhaps, by some French cook or landlord, to give the maximum of filling capacity and high-sounding *menu*, comprising a wonderful succession of soup, fish, flesh and fowl, ragouts and stews, followed by flabby puddings and other dyspeptic compounds, classed under "sweets," (a sop to Cerberus, the English) and ending up with knotty, wilted and colicky fruits, for "dessert."

The man who can withstand the contagion of example and the feverish thirst induced by the gormandizing of so much spiced meat, for a mortal hour and a half, without adding thereto a bottle of the "wine of the country," or some other country,—is a better total abstainer than usually finds his way across the Atlantic.

To add to this, the drinking water abroad has long been discredited as to purity and wholesomeness. According to a witty Frenchman, it has "so tasted of sinners ever since the Flood" that not many of the natives can be induced to indulge in it—and yet the Lounger survived the rash attempt of quenching his thirst therewith—on several occasions.

Well, the Lounger will suppose, that having sojourned for many days already at the preferred hotel of your favored town for the time being,—so long, indeed, that you have exhausted all the flavors of its special antiquities or celebrities, and all the antique flavors of the hotel "table-d'hôte" as well, until, after daily sequence of promotion at said table, step by step you have risen to the very top—you conclude it is now due time to "step down and out" once for all. In short—you are ready for fresh fields and new table-d'hôtes—you are prepared to leave! Of this fact, and the train you propose to take, you are expected to give timely notice. You do not, however, settle your bill at "the office"—the "bureau." At least that is not the customary way. At your last meal before leaving, you ask your table-waiter for it. He brings it—and its length will astonish you! You may be reminded of the classic saying that "our sins are like the dragon's teeth scattered by Cadmus: when sown they rise up like armed men against us." Here, the ghostly shadows of everything you have eaten for the past week, every little indulgence of the appetite, reappear and confront like those dread spectres to guilty Richard on the eve of Bosworth—and, like those, insist that the score shall be cleared off, the reckoning paid!

If you find the indictment to be a true bill, you send it back with the round money to cash it on the salver.

When the waiter returns it to you receipted, with the change, then is the time for you to vary the monotony of proceeding by giving a little change to him; leaving on said salver as many francs or marks or guldens as your liberality will prompt, or your purse allow.

Then when you leave your room, your chamber-maid, your room-waiter, your "boots" and your baggage-porter will rally around you, each seizing some satchel, parcel or wrap, and escort you, as a guard of honor, down to your cab. You feel that it would be the crowning disappointment of their lives if you should fail to bestow a gratuity on each, after all their kindness. A little page stands at the door, who has opened it for you many times. Please do not have so treacherous a memory as to forget him, much less that other boy who attended the "lift," which we call the "elevator." You may think that inasmuch as you have paid for every possible thing under the sun that you have enjoyed in this hotel, and probably a few more—and then paid roundly beside for '*service*' as a special item in the bill, that this lets you out ! But they dont think so, and they show that their sensitive feelings are so acutely touched that you can scarce fail to "remember them." You had thought, as a matter of principle you wouldn't—but as a matter of practice you do. Public opinion counts for something—and the public opinion of this hotel is unanimously against your view of it. Then if you have any gratitude left, after all these assaults upon it, you will "remember" the *portier* handsomely—for he really has been of great service to you. Such a treasure the unsophisticated Lounger found first at his hotel in Paris—and begged to be informed as to what his real position was in the establishment. The reply came in good English: "I am the 'All Porter."

He had learned English by *sound* alone—but it was truer as he pronounced it. He was the all-inclusive epitome of hotel knowledge and usefulness.

So after remembering fairly the *portier* that he may remember you gratefully, you are ready to depart. All the attachés are drawn up again, as they were on your arrival. The landlord bows and smiles farewell (you have really paid him a big fee in that charge for '*service*' in the bill) and if you have made the rest feel happy, they all bow and smile and wish you a pleasant journey, and you roll off toward the station, feeling that you are a pretty good fellow on the whole, and how pleasant it is "to scatter plenty o'er a smiling land!" But while your heart is lighter thereby—just so, alas, is your pocket-book!

AT HOTEL TROMBETTA.

AT LAST we were in Italy! In half an hour of the mild boredom of subterranean railway transit, we had traversed the sequence of ten long years and eight miles of solid rock, that first great bore of the Alps, the Mt. Cenis Tunnel.

Then, dashing down the gorges, through succession of a dozen other tunnels, great and small, with glimpses of cloud-wreathed, splintered peaks and more distant snow-capped Alpine summits, we sped along—following the picturesque course of the Dora—past villages and towns perched amid rocky defiles, with little checker-patches of yellow wheat, clinging to bare cliffs far overhead; on down through the chestnut forests of the foot-hills, and out to the fair plains of Piedmont, bright in the afternoon sunshine, and rich with grass and fruit and waving grain.

We stayed for the moment our steps in Turin, that city of ancient lineage but of modern glories; the chief being that it was the birth-place of Italian Unity, and that it haply nurtured King and Statesman, whose patriotic ambition refused to be constrained within the narrow confines of the old duchy of Savoy. Here in Piedmont, but a brief generation ago, lingered the sole remnant of Italian independence, the last shadow of nationality for that race whose ancestors once dominated the world. From Turin, as a center of influence and action, radiated the inspiring hope, the far-reaching thought and plan that compassed successively and finally, throughout all Italy, the overthrow of hateful tyranny, the expulsion of the Austrian and the Bourbon. Then came the moulding of all those diverse states into one autonomy, the realization of that patriotic dream of a United Italy.

The scepter and sword of Victor Emanuel, the musket of Garibaldi, the restless heart and fiery tongue and pen of Mazzini; these were, indeed, potent factors toward triumphant success,—but what had been all these but for the repressing, controlling, commanding intellect, the mighty brain of Cavour!

It was a most appropriate sequence that, for six years, Turin should enjoy the honor of the seat of government, the capital and court of the new Kingdom of Italy. Then she gracefully ceded it in turn to Florence and to Rome.

We quartered ourselves at the Hotel Trombetta in the Via di Roma. This hostelry is, perchance, not "starred" in Bædeker but, as memorably happy days are sometimes said to be "set with a white stone," so this first evening in Italy shines out now against the darkness of the past as a white star, in conjunction with the young comet which actually heralded the Lounger's advent—and some of its mild effulgence illuminates in recollection even the commonplace Hotel Trombetta. We had heard much of the overreaching nature of the Italian landlord, but by mine host at Turin, our first example, if the Lounger was in any wise defrauded he never discovered the fact, and so in any event, according to good authority, "was not robbed at all." On the contrary, we found him, whether in or out of that little den of an office on the ground-floor, most courteous and obliging—with his hotel all one could reasonably ask. The tourist-season was now nearly over, there were but few guests in the house, so we had it mostly to ourselves, and the best of it, including a room on the first-floor, opening on a gallery, which in turn looked down upon an inner court below, away from the noises of the street.

After removing the soil of travel and refreshing ourselves with supper, we found it still early enough to make some survey of the city and surroundings,—so the Lounger queried of the landlord whether it would not be well to take a drive. The jolly host not only spoke English fluently, but replied in graceful diction, which, perchance, in the mellow cadences of his native tongue might have melted into poetry itself:

"Why not, Signore? Is there not ample time? 'Tis indeed the favored hour. The sun has just gone down behind the Alps; in the beautiful public-gardens of the suburbs and along the banks of the Po, the air will be refreshing and every prospect at its best. At your service, Signore,—shall I call a carriage?"

This was indeed up to the mark of the Lounger's high anticipation that in this land of the south even the landlords would have a pleasing grace of expression. And yet he may possibly come to find before he leaves Italy that while some of them are given to romance, they do not invariably speak in blank verse!

So the carriage was called, and rolled down the Via Roma, past Piazza San Carlo, where stands in imperishable bronze the statue of Duke Emmanuel Philibert, old "*Tete de Fer*," mounted as when he rode into the fray at St. Quentin's, winning the battle over the French for Philip II. and regaining for himself and descendants the Duchy of Savoy. Then on to the more spacious Piazza Castello, the heart of the city, with its Palazzo Madama, the ancient castle of William of Monferrat, and the more modern Royal Palace, which exteriorly has little appearance of a palace at all.

Thence by the broad avenue of the Via di Po, flanked by arcades with handsome shops which seen later when fully illuminated for the evening, present a most brilliant appearance,—to the Po itself, and by its banks southward, a beautiful drive, surveying the heights opposite, the historic "hill of the Capuchins," with villas overlooking the city and country; and on to the New Public Gardens, whose extensive grounds include not only a Botanic Garden but also a royal chateau, of the seventeenth century, "Il Valentino," now occupied by the Polytechnic School.

And then we are driven across the city to the western suburbs, where we get a view of the Alps stretched against the evening sky, from Monte Rosa and other snow-crowned peaks of the Pennine chain on the north, sweeping the far circuit of the Graian and the Cottian Alps to the west, in the long arc of almost a semi-circle, around to Monte Viso on the southwest pointing down toward the Maritime Alps which, with the Appenines, shut off Piedmont from the sea. Grand seat in her fertile valleys and grand background against her snowy mountains has the beautiful city of Turin!

The return to the city and our hotel is made by way of the Piazza D'Armi, the grand course, where all the wealth and fashion of Turin display themselves and their fine equipages, at this hour of summer evening. The drive has been a varied and interesting one and, taking it all in all, the charge therefor fails to strike the Lounger as very exorbitant, being but two francs, or forty cents of our money, *pour boire* included!

And later, until the warm dusk of summer evening gave place to the freshening coolness of dewy summer night, we sat out upon the balcony of the hotel, and watched the throng in Via Roma below, as they sauntered along; the ladies with no other head-gear than a corner of black lace, but every man, woman and child carrying and lightly waving a fan.

So as we sat and noted the crowd, all apparently as light-hearted and merry as their favored race should be, our visions, like the last gleams that had faded an hour or two ago on the Cottian summits, were rose-colored. So far there had come to us no hint of disillusion or disappointment. The overture was just as it had musically been ringing in our ears since the days of boyhood. The curtain would now roll up and the opera (Italian) would begin. We were already within the happy confines of that land longed for during so many years. Land of sunny skies and flowery plains and fertile glebe that for more than two thousand years had yielded a free tribute of corn and oil, of fruit and wine! Land of gray history and great deeds of old! Blest land of art, of music and of song! We should come to behold its heaped-up treasures of the centuries past, its grand dower of Sculpture and Architecture, its pictured glories of the Renaissance. We should float in our gondola by moonlight, between rows of marble palaces, on the Grand Canal; we should come to lodge in a grim old mediæval fortress-palace of Florence!

And so we lingered before going to rest and more happy anticipatory dreams—and thanked the kindly stars that looked down upon that balcony of the Hotel Trombetta, that at last we were in Italy!

AT PENSION GIOTTI.

"FIRENZE," shouted the "guard," as he threw open the door of our compartment: "Fi-ren-ze," reiterated half a dozen railway porters, their eyes as well as their tongues "in a fine phrenzy rolling,"—and we stepped out upon the platform of the Stazione Centrale of Florence, —"Firenze la Bella."

We had just come into our long-coveted heritage of Italy! We had traversed the fertile valleys and fair plains of Piedmont, and then sped southward across the Appenines. From the crested heights of its bold semi-circular arc, mantled with bristling enceinture of frowning forts, and girdled with its adamantine wall of mountain, we had looked down upon a city, whose association with the name of our great discoverer has made it memorable to all Americans. We had noted with admiring eyes its rising palatial mansions of brightly painted marble ranged on every hand, substantial evidences of present prosperity—and then surveyed beneath, the narrow streets flanked with ancient palaces that still speak eloquently of the old-time glories and greatness of Genoa the Proud. Beyond the two Moles, which form an intermitted chord subtending this arc and enclosing the sightly harbor, spread out to the infinite horizon, sparkling and flashing in the noonday sun of the Levant, we had beheld, for the first time, the beautiful Mediterranean, and like the soldiers of Xenophon, we had exclaimed, "the Sea—the Sea!"

Again with the speed of the express, we had stretched our course along the never-to-be-forgotten Riviera, one of the most picturesque of railway journeys in the world. On our left, the bold cliffs, or sunny slopes of the Appenines, clad with forests of wild olive, or dotted with bright villas owning fruitful groves and shining with tropic luxuriance of lemon, of aloe and of olive; with anon, rugged and ragged walls of moldering old castles, perched on rocky heights, or the opening clefts of valleys coursed by brawling streams tumbling down to the sea; or the breaks filled in with a heterogenous repletion of irregular old towns and villages, made up of tall and narrow old stone houses piled up in apparent promiscuous confusion. And on the right hand, always the broad expanse of the Mediterranean, bright in the sunshine, blue in the shadow, as the clouds floated lightly away, or as lightly hung poised in the heavens.

At one moment we had skirted a long sandy beach, or had rounded one of those still bays of peacock-blue; at another, the track pierced a bold promontory, and our train had dashed into the cool depths of rock, where through lateral galleries we caught flashing glimpses of the white of sky and the blue of water—and the next moment we would be again out amid the dazzling sunlight, in full view of mountain and of sea. Some eighty tunnels repeat for four-score times this shifting play of color, this alternate transition of the darkness and the light.

On past Spezia, with its beautiful bay now forever associated sadly with the memory of the gifted, the unfortunate Shelley; on past many an old mediæval town and fort, or the ruined site of some far older town of the days of the Romans,—and there rises at length before the traveler, a vision of the beautiful Carrara Mountains, distinct in summit outline but gray, spectral, ethereal, poetic; the very wraith or ghost of a mountain range, save, indeed, where in spots gashed and riven by quarries for their renowned marbles, they shine dazzling white in the sunlight.

Then pausing at Pisa, the Lounger had fairly twisted the axis of his eyes in gazing upon its wonderful Leaning Tower, the puzzle of the centuries. Much idle speculation has been indulged in, indeed, concerning its deflection from the "*Perpendicular*" style of architecture—overlooking the obvious explanation that its old contractor-builder, growing dubious when half done as to his final outcome of payment, had decided to take a lean upon it thenceforth! The Italians of those times even were well versed in finance—the Lombards of Italy far ante-dating those of Kansas City. Building operations were well comprehended—by the Banks and Loan Companies—and fine blocks went up, fitted out with the "modern improvements." Many a sumptuous palace had a mortgage attachment—and nearly every Campanile in Italy took a lean, as is apparent to this day. That was the reason they always detached the bell-tower from the Cathedral, so that the latter wouldn't have a lean-to!

Grateful in the hot mid-day had been the marble Baptistery of Pisa, cool and dim—and in its venerable Cathedral close at hand, the Lounger had viewed with interest that famous old bronze chandelier, whose vibrations once happily suggested to Galileo the first idea of the pendulum. Unfortunately, when it came to applying these discoveries of scientific principles, his life never "went on like clock-work" thereafter.

And then, three hours' ride up the valley of the Arno,

through the rich heart of Tuscany, had brought us at last to Florence. Historic, artistic, poetic, beautiful Florence! At the mention of thy name what memories throng our minds! Swift retrospect of factious fighting Guelphs and Ghibellines—of plotting Medici, and Machiavelli; admiring recollection of Dante and Boccacio, of Galileo and Alfieri; fond associations with Cimabue and Giotto, with Filippo Lippi, Botticelli, Luca della Robbia and Donatello; grand, inspiring memories of Fra Angelico and Savonarola, of Brunelleschi, Ghiberti and Michael Angelo!

* * * * * * * * *

Has the Lounger left the reader too long at the railway station in Florence—or rather, carried him too far back upon the road! Well, then, we will at once take a cab, of which there is a fair supply on hand. It is but a short drive to the center of the city, and immediately we pass one of the notable churches of Florence, San Maria di Novello, dating back to the thirteenth century, and boasting the celebrated frescoes of Ghirlandajo. Striking into the Via dei Fossa, we are soon on the banks of the Arno. One look at this long, straight, turbid canal, diagonally intersecting the city, is sufficient—the romance is all gone off the Arno!

We cross the Ponte Carraja, for we are bound for the south side, "Altr' Arno." The recommendation of a friend has committed us to a new experience, we are to be quartered at a *Pension*—"Pension Giotti." This broad quay of a street that lines the Arno on that side is Lung D' Arno Soderini, and the grim old corner building ranging at the head and extending down the thoroughfare of Via Serraglia is the old Soderini Palace and our destination. Now comes the exciting part of our drive.

The Italian Jehu can wield the whip-cord even as the cow-boy of our plains, and with a perfect tumult of lashes and cries and a grand fusillade of whip-crackings, we storm down the street to the portal of Pension Giotti! This is the conventional arrival. It is, indeed, no more a martial cavalcade of mediæval chiefs in glistening mail, with a retinue of bold men-at-arms, coming to claim the hospitality of the Palace; it is not even any of the heads of the great House of Soderini themselves, returning home from council, or the field! Notwithstanding all the din, it is simply a pair of humble Americans in a "one-horse shay" of a *fiacre*, coming to quarter with Madam Giotti in her Pension at eight francs per diem, "wine of the country" included!

Warned by the signal, ancient Francesco—major-domo, head-waiter, porter, and general factotum of the establishment, the surviving Caleb Balderstone of the Palace—appears to receive us, and we are ushered up the old staircase to the first-floor, where we are welcomed, in good fluent English, by Madam Giotti herself. She is a pleasant-looking landlady, fat and fully forty, though, being Italian, hardly "fair" as well. As to Mr. Giotti, Signore Giotti, if, indeed, he existed in the flesh, we could have seen him but once, at day of our departure, associating him then with personality of the man that made out our bill. Unlike Madam, he was scant in rotundity of figure, and therefore not at all "round as the O of Giotto." In extent, indeed, somewhat like his own bills, he was in all respects "a spare man."

We are assigned a spacious, carpeted room adjoining the salon and overlooking the Via Seraglia. This outlook is pleasant by day, whilst most of the old building otherwise is rather gloomy and dismal, with its brick and stone floors and its dingy corridors. Portions of it impressed us as somewhat on the order of a states-prison for political offenders. The guests just then were but few, only some four in all besides ourselves, and all English,—chief among whom in the Lounger's recollection is a literary-minded young man of delicate lungs, and his widowed mother, evidently devoted to his care. The preceding winter they had spent in Corfu, and soon they would be taking flight, to the cooler shades of Vallambrosa.

The Lounger gleaned only a partial history of the notable building in which he was lodged. It dates back into the Middle Ages. In one room connected with the pile, Niccolo Soderini received St. Catherine of Sienna, some time about 1380. But of this noble family, the most illustrious was Piero Soderini, who at the beginning of the sixteenth century was given life tenure as chief of the commonwealth, "perpetual" Gonfaloniere of Florence. The history of his administration is intimately connected with that of Machiavelli, who served under him, both in important embassies and as conqueror of Pisa. By force of his great ability usually influencing and swaying largely the Soderini, yet Machiavelli could never quite forgive that it had been impossible to bend him wholly to the exercise of that subtle or strenuous policy which he himself alternately favored. In those wicked times, the crafty secretary could hardly account for such simplicity of character or goodness of heart, contemptuously satirizing them after the Gonfaloniere's death, in these lines:

"The night that Peter Soderini died,
  His soul flew down into the mouth of hell;
'What? Hell for you? You silly spirit!' cried
  The fiend: 'Your place is where the babies dwell!'"

But the old Palace has seen its best days. Its ground-floor seemed even more lonesome than the *pension* above. The Lounger recalls a cigar-shop, indeed, and up toward the bridge, offices of some kind; possibly a bank, a broker's, a barber's, or some other form of shaving-shop— being inconspicuous enough for either. Whilst but little commerce was carried on in the old building, it was just that much more than owned by its next neighbor, the Rinuccini Palace at the corner of Fondaccio di San Spirito. This more modern edifice, with its coat-of-arms over the doorway, had been built in the sixteenth century, by that painter, sculptor and architect, Luigi Cardi Cigola.

But across the way, on the opposite side of Via Serraglia, there were more signs of activity. First, a wine-shop on the corner; next a green-grocer's with a luscious store of ripe tomatoes, and then came a fruit stand, just opposite our windows. Some of the tomatoes the Lounger was daring enough afterward to buy and send with his compliments, by Francesco, to Madam Giotti, with request that they might be sliced down to add variety to the regulation dinner. Following this measure, the Lounger's plate was thereafter abundantly supplied with that "esculent." No doubt the Italians themselves delight in vegetables, for their markets abound with them—but, as the regulation table-d'hôte dinners are confined to fish, flesh and fowl, they no doubt opine that the foreigner desires none other.

And the Lounger yet hears, in fancy, the shrill note of that fruit vendor opposite, as in the hot stillness of the lazy afternoon he would break out in a spasm of "*Ci-li-geio!*— *Bel-la! Bel-la!— Ci-li-geio!*" Cherries— beautiful cherries!—and sure, the cherries of Italy in their glossy red or black pulpiness, are both beautiful and delicious; but all the livelong day, and far into the hot summer night, we got tired of hearing of it, all the same! And yet other noises there were that, one or another, arose to the Lounger's window, and in conjunction with other matters, "murdered sleep" for him! The shuffle and tramp and clatter of many feet along the middle of the street! The sidewalks in many of the towns in Southern Europe being largely devoted to other purposes—as other Loungers abroad will be reminded—much of the foot-travel takes the middle of the street as, on the whole, least objectionable. Then the late customers at the corner wine-shop below would get boozy, and, judging by the outcry, were declaring in obstreperous, drunken Italian that they "wouldn't go home till morning!" Then the cabs and the carriages returning from the Cascine, and every other place in and around Florence, all seemed to strike Via Serraglia, the great thoroughfare from one side of the Arno to the other, across the city. Hour after hour they would throng noisily by. It did seem as if the Florentines never went to bed on summer nights—especially on moonlight or *"festa"* nights!

At last, even the very latest pair of noisy, boozy, quarrelsome, wineshop customers had departed. The sound of *"Ci-le-geio bella"* had long since been hushed. A single, minute representative of that acrobatic and elusive tribe of Southern Europe, which has been characterized as "the wicked," even "when no man pursueth," even that "wicked had ceased from troubling." Silence had finally descended "like a poultice to heal the blows of sound." The weary and exhausted Lounger would at last be sinking into rest. But just before this final consummation of unconsciousness, there would come a far-off echo, a faint rumble which would gather into a more decided ryhthmic roll. And this would grow, and louder grow, "nearer, clearer, deadlier than before," till finally, with clatter and crash and whip-cracking and yell, with all the vigor of Pandemonium let loose, once more down the Via Serraglia would come the Italian Jehu with his *fiacre*—in front of the old Palace Soderini; in front of Pension Giotti!

AT SUMMIT OF SIMPLON PASS.

WE STAND uplifted to the skies!
But higher yet the mountains rise,
Above, around, on every hand;—
With range to our range opposed
Th' vast horizon north is closed;
At will the eye may rove and turn,
From Pennine chain to Alps of Berne:
Across a mighty chasm, stand
In serried rank, serene and grand,
The shining peaks of Oberland!

Encircling mountains by the score,
Whose thund'rous avalanches shower
Down mighty slopes, with Titan power,
The snows that feed its swelling tide,—
Rifted and wrinkled and petrified,
The Aletsch glacier's torrent vast,
For many a spreading league is cast.

A part of its sinuous course is hid
Behind black cone and pyramid,
That, jutting skyward, appear to be
Tall islands in a frozen sea!

Ever the glacier holds its course,
Type of the slowest, surest force!
Ever the Aletsch swells and gains,
Bearing on to its grand moraines,
And surging downward, to depths below,
Not only flows, but seems to flow.

* * * * * *

Turn we at length from fields of air,—
From fields of ice, and summits bare,
With winter reigning everywhere!
Here, miles away, yet at our feet,
There lies a prospect fair and sweet;
Sprinkled with dots of color warm,
Of town and chalet, croft and farm:—
With thread of silver seaming down
A verdant vale, that stretches on,
Hemmed in by giant walls of stone,
By Brieg—by Leuk—past far Sion,—
The long green valley of the Rhone!

SOME SWISS INNS.

SOMEBODY has described Switzerland as a land of mountains and hotels, and certainly, within the circuit of its confines, the innkeeper may be regarded as almost ubiquitous. If you ascend to the top of the mountain, lo, he is there! If you make your bed in the remotest valley, the very next night he will be there to make it for you, with fairly clean sheets, for five francs,—"attendance" two francs extra. If you take the wings of the morning (the cheapest mode of travel in Switzerland, where railway "Passes" are scarce) and fly to the uttermost parts of Canton Grisons, he will already be on hand, with bread and butter (or honey) and a cup of coffee, for breakfast—at two francs!

In the wildest solitudes of the crags, or threading the slippery edge of a glacier, you turn a corner, and behold the Swiss hotel right before you,—with a smiling landlord and a knot of guests, who have already secured the rooms with the best view; each flourishing an alpenstock, with the names of a dozen Swiss passes branded thereon.

That venerable Alpine resident who once warned "Excelsior," in the language of the high railway-official when the "Inter-State Commerce" went into effect— "Try not the Pass"—has since been superseded.

You leave your restful inn at Chamounix, and the agreeable diversion of accompanying the tourist ascent of Mt. Blanc (by telescope), and join the long procession that toils up Montanvert. At the first halting place you discern a *cabaret* by the roadside with the inscription: "*Ici on voit un chamois, vrai et vivant!*" Surprised to learn that notwithstanding, or on account of, the plentitude of their skins in American drug-stores, a "genuine live chamois" has become almost as rare in Europe as the mountain-sheep of the Rockies, you stop to inspect the curiosity—and patronize the owner. As you progress farther up the woodland trail, other kindly natives start forth with proffers of refreshment for the worn traveler, and their spring is associated with this injunction:

"Drink weary pilgrim, drink—and pay!"

Arrived at the summit you find an inn, with a Parisian *menu*, and a lot of people eating and drinking therein. They are always eating and drinking in Switzerland—when they are not climbing mountains, or branding alpenstocks, or buying souvenirs of wood-carving and agate jewelry. It is a perpetual picnic wherever you go. Whenever the tourist gets tired of nature, or eating, he rests up by shopping for souvenirs—and the shops, like the inns, are always at hand.

You descend to the Mer-de-Glace, cross it, and, traversing the lateral moraine at side of Aiguille du Dru, your guide suddenly gesticulates and exclaims: "*Monsieur, voila le Chapeau, et voici le Mauvais Pas!*" and then plunges down the precipice, apparently. You follow him perforce, and when at length, at peril of your life you have rounded the last slippery point of the narrow, hanging shelf, you come to "the Chapeau"—where you find another lot of people, eating and drinking!

In the wildest haunts of nature, amid the deepest recesses of her mountains, you can always secure what you wish—if you will only pay. All the luxuries in or out of season, the choicest vintages and viands, the very best cuts of meat! The Lounger once heard an indignant Briton exclaiming at what he called the arrant humbuggery of the landlords and their French cooks. "A *filet*, you know, is a tenderloin steak. Well, these fellows always have it, or will provide it for you. One of them actually said to me: 'we are out, but for three francs I will make you a *filet*.' Just think of it, they would *make* me a *filet!* They would take a piece of 'round' and pound and braise and lard and spice and garnish it, and turn me out a *filet*."

In the innocence of his heart, this honest islander had always supposed that a tenderloin steak—like the poet—was "born not made."

And yet, notwithstanding the proverbial sordidness of the landlords, the theatric effects often imposed upon the scenery, and the "posing" of the tourists themselves, the Lounger found still some nature unspoiled, some grandeur and beauty yet unvulgarized, some inns enjoyable and worthy a recollection, even in Switzerland! Sweet was our rest "at evening's close," after the long day's journey crossing the Simplon,—at the homely hostelry of "Three Crowns and Post" of Brieg. Not at all to be despised was the hospitality of Hotel Clerc at Martigny, near where lingers yet that "old round tower of other days" which looks out upon snowy mountain tops and up the beautiful, green Rhone valley, hemmed in, on either hand, by mountain walls. No word of proud scorn has the Lounger for even the famed Schweitzerhof at Lucerne, with its band of music at table-d'hôte, its lively waiters, and its good round prices, which were also "square," with no extra "*service*" in them. Memorable shall be the humble "Lion Noir" at Altorf, the famous little village where, once on a time, bold William Tell shot an apple off his son's head, straight down the street at eighty paces—or around the corner as the street now winds, one hundred and twenty yards as the Lounger paced it. Whatever the actual truth in "the Tell myth,"—whether happening in Norway, in Persia or nowhere except in the imagination of all the Aryan race,—there was at least no mistake about the excellent quality of the lunch at the "Lion Noir," or the excellent serving by the respected daughters of "mine host,"—and their refusal of a fee afterward. This actually happened in Switzerland!

Neither does the Lounger harbor any harsh feelings toward the landlord of the Hotel de Londres at Chamonix, in that he raised the price of a room two francs on account of its superior view of Mont Blanc. Not that the Monarch of Mountains might not usually be seen just as well outside, and almost as plainly as the rushing Arve, soapy-gray with its dissolved detritus of glacier, and roaring by like a torrent beneath our window. But, late one night, from said window, when the Lounger chanced to wake and look forth, he received a revelation of the glory of Mt. Blanc! There, as it lay, still and silent under the moonlight, the light of a gibbous moon which swung low above, it had an air, not so much of majesty, as of serene, unapproachable isolation. It seemed so near, and yet so far. So near, as a glistening snow-bank on a neighboring hill-top, from the farm-house window of boyhood, in the midnight of winter night, crisp and cold. So near as a powerful telescope brings into the field of vision the clear disc of a planet—and yet, in imaginative impression, remote as lone Uranus, the weird specter of a world, swaying and drifting afar off, through the illimitable ether of space.

IN COMPANY.

It was one of the happy accidents of travel that brought into contact, at the same table-d'hôte at the old Hotel Victoria at Venice, the Lounger pair and a duet of ladies, independently taking their own "ways abroad," but who had become just a little tired of the loneliness of the same;—Fraulein Meister and Senora Grande. These ladies had just arrived from Germany and Austria, by way of Trieste. It was also a fortunate incident, as both parties professed on speedy acquaintanceship, that, on comparing respective routes for the next few weeks, these were found to approximate so closely that they might easily be made to coincide. Being well pleased with each other, these "strangers in a strange land" resolved, therefore, to form a temporary association, a limited copartnership of travel, which Company was incontinently dubbed "the Lounger-Meister-Grande Combination,"—and soon voted (by themselves) to be altogether the most harmonious and successful traveling troupe starring it abroad that season.

Possessing many like tastes and sympathies, each supplied to the common capital and stock in trade something desirable to the other. The Fraulein added good generalship and excellent German to the Lounger's inefficiency and execrable French. Senora Grande contributed a fine practical knowledge of railways, as well as supplemented the deficiencies of Bædeker by a most intimate acquaintance with the text of Mark Twain's travels. The equipment of the Combination, therefore, was quite complete. Together, in sable-painted gondola, by mystic moonlight or by garish day, they explored the canals and lagoons of Venice, her old palaces and prisons, her picture-galleries and cathedrals; together they exploited all the shops on the Piazza St. Marks and closed out their contents at fifty cents on the dollar—of what was asked. Together, on "taking the road," they traversed Northern Italy, made the round of her beautiful lakes and afterward crossed the barrier of the Alps by Napoleon's grand road up the Simplon Pass. Then, as a merry quadrille, together they executed that zig-zag dance, the tour of Switzerland—of which these are some of the familiar "figures"—"*Salute*," (at table-d'hôte)— "*forward four—back—forward again—cross over—* (mountain) *chain—dos-a-dos* (in diligence)—*swing corners*"—and so forth and so on to the end of the dance; when the final word was given, "*promenade all*," at Berne; and the Meister-Grande couple took their seats— in a railway compartment—for Paris, leaving the somewhat forlorn Loungers to stray around the hall, and finally across to Germany.

This memorable combination at all times abounded with two things of similar sound which should really be reciprocal—good humor and good-humor! If the wit of the former was sometimes at the expense of the Lounger, he trusts that he never forgot to keep himself in the latter. But outside of this, he may possibly have discovered some trifling disadvantages which at times placed him on the debtor side of the firm's books. These books were certainly out of balance in one particular—he was always in a hopeless minority as to sex; and what man can properly maintain the dignity of his estate when in the proportion of only one to three! Any system of minority representation proves of but little avail under such circumstances. First and foremost,—when thus backed by such a force,—the Queen Consort, who, temporarily, through her ignorance in respect to the *menu*, the speech and the "wine of the country," had been reduced to a reasonable state of domestic restraint and subjection to the powers that be, now suddenly threw off the yoke and demonstrated that we shall all be changed. In the twinkling of an eye, great "I" was obscured—put out. The Lounger was summarily deposed —and Queen Consort reigned in his stead!

Thereafter he was no longer first in a feast, or last in a fray—of discussion! In fact the whole method of reasoning was changed, or rather, intuition superseded reason entirely. Things future were prognosticated from the stand-point of infallible insight. Things present were viewed from the vantage-ground of inner-consciousness. The tyrannous majority looked at all things subjectively— and it was no manner of use for the Lounger to treat them objectively. It was idle for him to put himself in the objective case, when they were always in the possessive and nominative. Or to interpose a subjective "if" when they were plainly in both the indicative and the potential mood—and if necessary would be in the imperative as well!

Then when it came to the consideration of the past, they would always remember things and places associatively. The reader will recall a story illustrating how ladies illustrate their travels. A lady and her daughter had spent a year in Europe. On their return home, a morning caller asked, among other chat—"and you visited Rome I suppose?" "Rome?—Rome?" hesitated the mother—"Fanny, were we in Rome?" "Oh, yes, Ma! Why, dont you recollect! Why, that was the place we got the bad stockings." And then the mother recalled Rome—through association!

Even yet the Lounger can see Fraulein Meister glancing down meditatively and approvingly on a pair of well-wearing (not well-worn) gloves. "Oh, we did have such good times in Dresden! It is such a nice town. You must go to Hotel de Saxe, they have such a good *portier* there. He knows everything and was so accommodating. He told us of all the best places for Saxony lace, splendid Duchess and real Point, and so cheap! I'll give you the addresses, and also of that little shop where I bought these gloves. I wish I had got more of them. Such a nice girl sold them to me. It was a little shop not far from the Zwingler, where the Gallery is and the Sistine, you know. There was nothing else in the room at all, and such divine eyes, so mystical and yet so tender! They have certainly worn wonderfully well—but it is about time for a new pair though!"

At the end of this monologue, the Lounger begged that he might have the separate addresses of the shop, the shop-woman, the Dresden gallery, the Madonna, the eyes, and the gloves—for, somehow, they had all got badly mixed up in his mind! But the ladies unanimously avowed that it was all plain enough already—only these stupid men never could get things straight—by intuition!

It is associatively too, that these ladies recall to this day, that little, old, double-barreled, Franco-German, bilingual town of Freiburg. Here the boundary line is distinct between the two spoken tongues in the same town. A "great gulf is fixed"—the deep gorge of the river Sarine—but traversed by a high suspension-bridge, which unites them. It is one of the quaint, picturesque and wonderful towns of Switzerland; and it contains one of the musical triumphs of the age, the great Freiburg organ with its wonderful *vox humana* stop, which strangers come from all over the world to hear. And yet, if you ask the ladies about Freiburg, they will tell you they recall it well, for in the cosy parlor of the Hotel de Freiburg they discovered the first pair of rocking-chairs—yes, real American rocking-chairs—that they had found in all Europe! Now the Lounger will say in justice to these "*compagnons de voyage*," that they were usually amiable (except to him) and self-denying, "in honor preferring one another." They would sometimes proffer even to yield up to each other the better room in hotels, and that is going a great way in self-sacrifice for fellow travelers. But when they saw those "rockers," they stood on no sort of ceremony. There were but two of those long regretted luxuries of life—and one lady got left!

A HOTEL ROMANCE.

LATE one summer afternoon, the Lounger was strolling down the Quai de Mont Blanc at Geneva, accompanied by his traveling-troupe, a bevy of three ladies. Be it known that the logical Lounger deems that he has as good authority to apply this term "bevy" to his summer companions as hath the ornithological editor of the Journal when describing his "Winter Companions" with "their flitting and twittering." There may be, however, a distinction, with a difference! Our younger student of animated nature is able to know his subject exhaustively, as well as to set forth faithfully and interestingly every varied characteristic,—while the Lounger is forced to confess humbly, with regard to his, that their ways are oftentimes past finding out!

While thus strolling along, contemplating the blue waters of Lake Geneva, and striving to pierce, in the southern horizon, the baffling haze that veiled the distant view of Mont Blanc, we suddenly and most unexpectedly came right upon a hotel romance! At least, it happened at a hotel—and the ladies insisted that it was nothing less than a genuine romance.

Just as we arrived opposite and in full view of "Hotel de la Paix," we espied a young lady fair sitting out alone on a little hanging-balcony of its second story. Then at that precise moment, entered into the scene, by coming out on the balcony, a young gentleman. And the young lady rose up to greet him, and they kissed each other in the face and eyes—of all the people below, especially the ladies of the Lounger party! These were intensely interested at once, in the *denouement*, which they said was that of a veritable romance. Love on a balcony—a modern Romeo and Juliet affair, done in daylight! In vain the Lounger, being a family man himself, offered the commonplace but common-sense solution of the affair—that the fair one was simply a young married lady, boarding at the hotel, and now giving the appropriate conjugal salute (albeit rather public, indeed) to her husband just returning to her and his dinner after a hard day's work at "the bank" or "the office." The ladies wouldn't hear to this for a moment, declaring it was preposterous. On the face of it, they could see, even at that distance, that it wasn't that kind of a kiss at all. As their three pairs of eyes were sharper, as well as brighter than the Lounger's dull and failing orbs, he had to give in. Yes! it was an undoubted romance—and if the Lounger had the heart of a man in him, and the pen of a ready writer to command, he should write it all out and give it to the world. They would supply the (imaginative) facts! The couple we had seen were long parted lovers, beyond a question. The course of their love had not run altogether smoothly. On the contrary, it had been exceedingly rough for (as well as on) them. There had been a serious impediment—an obstruction,—

The Lounger here suggested—an obstinate old husband who had persistently refused to die off? But they wouldn't listen to this popular style—the Ouida-Saltus style of romance of to-day. Theirs was of the good old-fashioned sort. So, they scouted—No, indeed! A flinty-hearted parent, or a tyrannical old guardian, in connection with the poverty of the lover—this was the trouble that had interfered with the happy consummation of the attachment! So the maiden had been locked up at home, and the young man had gone off to seek distraction in travel. The kindly Lounger, willing to help out, again suggested—as a conductor of one of Cook's or Gaze's "personally conducted" parties—that would certainly afford him distraction enough!

At last the obstacle had been removed! (By arsenic, or a bill of divorcement!) No, no—dont interrupt—by the young gentleman coming into a fortune, or the obdurate rich uncle going out with an influenza;—and now, the lovers, separated by the breadth of Europe, had come together by appointment, to meet at this hotel in Geneva!

It was this happy reuniting and the sealing of their transport by a kiss, that we had just witnessed. It was the most fortunate coincidence in the world that we had happened along at the precise moment when the lover had arrived, looked at the hotel register (the Lounger here interposed that people didn't usually 'register' in Europe) —well, then, inquired of the *portier*—and then rushed up, and found her on the balcony! They wouldn't have missed the moment for the world. It was just too lovely for anything, and it was the very first romance they had come across in Europe! They had seen nothing in Italy equal to it. The Lounger felt impelled to remark that the young man seemed to be in pretty good style of evening-dress, considering he had just arrived from such a long journey.—Oh, well, of course, he had taken time to dress. You wouldn't have him appear unto his ladylove all covered with railway dust and grime, this hot weather! No, he had just waited for his trunk to come up, and in the meantime had a good bath in the hotel, and made himself presentable! The Lounger was glad, indeed, to learn that the young gentleman had got his bath so readily and easily! For himself, he had found it one of the hardest things to obtain at a continental hotel. He could recollect that at the largest hotel in Milan, with their one bath-tub, they had seemed disposed to economize time and hot water by bathing the guests in pairs.

Now as the Lounger confesses that he can scarce construct a romance out of such thin material—the incident being but slight, and the ladies' facts for a foundation somewhat problematical at best,—he would prefer to turn over the contract to some one else; to sub-let, in fact. To kind Mrs. Gray, of ladies-luncheon fame, for instance: she who once inquired, who was this maundering Lounger anyway, and why dont he write something interesting?

Here it is Mrs. Gray, and all our ladies said it was intensely interesting, and one of the most inspiring spectacles they had seen abroad; happening right there in Geneva too, where so many spectacles were made. Take it, and weave around it the web of romance! Or paint us the picture—if not in words, then in color! The canvas, at least, is a good one. Brush in the Quay de Mont Blanc, with our group of admiring spectators, disposed just alongside of Lake Leman with its tinted delicious blue. The blues will work in nicely. And then, only a little distance to the right, the "arrowy Rhone" just issuing from the lake, on its journey to the sea. The Rhone, a trifle "off color" in its fast and furious and decidedly improper haste to get there. Across the stream, and in the middle distance, the rising heights of Geneva! Paint in what is left of the old city-wall and the new watch-towers of the watch factories. Above all, for it stands above them in fact as well as in historic recollections, paint in the old house of John Calvin— and the memory of Calvin himself. Oh! You haven't enough carmine for that! Well then, save some for this hill beyond—the "Place de Champel"—where Calvin put Servetus in the way of being roasted, in order that his heart might "within him burn to know the better way." It puts one out of all patience with heresy, to recall that in spite of Calvin's kindly efforts to convert him, Servetus obstinately refused to turn from the error of his ways—even when done to a turn!

Out there in the far perspective, indicate, oh artist, as fairly as may be, the vision of Mont Blanc! Mont Blanc, indeed, can hardly be seen every day, or at every hour perhaps of any day—but the picture of Geneva without the distant view of Mont Blanc would be out of all character if not "out of drawing." So it "must go in."

And now, work up the foreground! Put a good deal of strength into the balcony, for, slight as it is, it must safely hold two enraptured lovers, disregardful alike of time, space, spectators, and the tenacity of the iron brackets beneath them. Then introduce the lovers themselves—or rather, as they have long since been "introduced" of course, paint in, first the lady with a very engaging expression on her face, and then work in the hero with an "engaged" expression all about him!

Last of all,—the kiss by all means!—

"Jenny kissed him when they met,
Jumping from the chair she sat in;
Time, you thief, who loves to get
Sweets upon your list, put that in!"

And Mrs. Gray, do not fail to imitate old Time in this one respect, and—put that in!

THE OLD BIBLE INN.

QUAINT old hotel, in a delightfully quaint old city, is the venerable Bible House of Amsterdam.

Approaching the city from the sea, the traveler enters the "Y," an arm of Zuyder Zee, which thrusts its wrist of a harbor and its fingers of canals into the quaggy land. Reverse the approach and the application of the figure, and one might liken Amsterdam itself, from the similitude it bears, to the outline of a human hand from wrist to finger tips, with its lines of tendons and network of arteries and veins. It is the broad, thick hand of the Hollander!

Six centuries ago, the original Dutchman came here where the sluggish, muddy Amstel slowly filtered into the sea. He put down that hand, strong and broad if somewhat pudgy withal, making its impress upon the peat-bog which was all that then represented dry land,—and the quaking bog became fixed and solid.

With all the assiduity of a beaver, he sharpened timbers and drove down piles, any number of them, yea piles of them. Be it not forgotten that Amsterdam bears a name as legitimately derived as that of Beaver Dam, in Wisconsin, or the more profane-sounding Yuba Dam, of California. So, like the beaver, he built the Dam of the Amstel. This was partly in the nature of a prohibitory amendment, to shut off "the drink" of Zuyder—Zee? Likewise, by diking all along the Amstel, he set up a bank. Thus getting solid footing, he put his foot down to hold his ground, and stay. Moreover, being of an investing turn, he resolved to put in all his pile here and that nothing should break that bank. Then he built more dikes and dams, and constructed bridges—three hundred of them,—and drove more piles,—"three hundred thousand more," or less. Some ninety islands were enclosed by canals which he had led around the plat in concentric circles, or strung across the lot, until his ground (and water) plan resembled a gigantic spider-web. And thus he literally founded a city. So much for Dutch perseverance! Touching a tender spot in the heart of Holland, getting hold of a soft thing in land, instead of Lady-Macbeth-like weakly crying, "Out, damned spot,"—he went ahead and dammed the spot in good earnest, and stayed it until it stayed in, permanently. Apparently there had never been the slightest foundation for the idea of a city here until this phlegmatic Dutchman went to work and put it in!

And his successors, too, by constraint of the great laws of heredity and necessity, were long kept on the drive to make a living, keep up the boom, and keep out the sea. They kept on building and driving piles, and now the foundation mud of Amsterdam has "millions in it"—so to speak. That is, speaking roughly and roundly.

Whenever they wished to "keep a thing in mind" or a house on a lot, they remembered the old injunction, to "stick a pin there"—and they stuck one in so multitudinously that there is now no lack of "underpinning." They had heard that it was necessary to "drive their business"—which was to pile up the sea—or the sea "would drive them" out, so they made it a matter of public duty to work instead of vote for "protection." It was only factious, narrow-minded dry-goods dealers who persisted in getting up "special drives" of their own.

When they came to build the Stadhuys Palace, they set it up on no less than 13,659 piles. Some reader may suspect that this is piling it on pretty heavy, and it pains the Lounger exceedingly to be thus definitely statistical, but this is the exact net figure—no discount for cash. Like the Ark that was pitched within and without, and had a roof of double-pitch, besides, this old Palace, both above and below, is the exemplification of a stately pile. This is "on the square"—of the Great Dam,—and only a short distance from the Old Bible Hotel, which the Lounger will get back to, by way (rather devious, indeed,) of Warmoes Straat. But the convex curvature of the street itself is as nothing compared to the deflections from perpendicular uprightness of the buildings that line it. This is not at all peculiar either, to this one street or city. Almost any Dutch town whose family name is Dam can boast buildings as various and varied in verticals as the campaniles of Italy. This interferes sadly with rectitude of alignment. There is no consistency about them, either horizontally or vertically, and as to the latter, the "section lines" of any adjoining pair of fronts, having but one single point of contact, cross and assume the appearance of the letter X.

At last we arrive at the Inn, an ancient structure, dating back to the period of the Reformation, when it was used for a printing-house, memorable especially for its early issues of the Reformed Bible. It displays a Bible for a sign and still preserves, as a valued relic, a copy issued from that old press of three centuries ago, the first Bible printed in Holland. The Lounger will scarce print any Commentaries upon the black-letter text of this venerable volume, for verily it was "all Dutch to him!"

Inside the little hotel everything is very comfortable, though on a small scale and contracted throughout; with such dark and devious halls, narrow corridors, and steep staircases as one might expect to find about an old printing-office. There lingers now, however, little else to remind one of that old black-art. The Inn boasts no parlor, and the little reading-room, fronting to the canal in the rear, serves for breakfast and smoking-room as well. Besides this, it is the "office" of the head-waiter to keep his accounts. And yet, it has its charm, both of quaintness inside, and of view from its windows, that open upon the canal, the Dam Rack, which is, in one sense, a leading thoroughfare of Amsterdam, being the widest of its canals, as, in fact, the course of the Amstel River itself. This view offers us constant temptation and enjoyment. The current of life and traffic moves lazily and slowly though methodically here, as does the outward current of the Amstel, or the sluggish tides setting in through the "Y" from the Zuyder Zee. Occasionally the little steamer that plies around the harbor and to the little neighboring islands, comes puffing in, and slowly discharges its freight or a few passengers on the landing almost opposite.

And then the old buildings and warehouses that line the canal on the other side are a great source of interest and speculation with us. Some of them appear so evidently built for business and to own so little of it. Each tall and narrow, and with its antiquated gable to the street, but all of varied architecture, they agree in little but this, that every one projects that gable, up at the cone, into a little peaked dormer wherein is fixed the necessary pulley and tackle wherewith to lift merchandise from the canal below up into the storehouse. But the merchandise that should enter or depart from them never seemed to materialize while we were around. Possibly the traffic had departed to more favored quarters.

Everywhere, indeed, in Amsterdam we were really delighted with the street views; they exhibited so much charming irregularity, and they revelled so fully in the elements of the picturesque, in varied form and color, that we revelled in them, in turn. Fantastic old gables, richly-red or mellowed facings, tiled roofings, turrets and towers of all forms and fashions; broken skylines above, shadowing trees and reflecting water below; these, with a "misty-moisty" atmosphere, combine to afford the *materiel*, and form the *milieu* with and in which the Dutch painter has demonstrated his ability to deal and work most effectively.

By the way, the Lounger understands that the Dutch water-color artists begin construction of a picture on similar lines to those on which their picturesque Amsterdam was constructed; viz:—they first suffuse their base (the paper) with water! After this they lay on and build up the color, and thereby get as soft a texture and as efficient a blending as their own moist atmosphere imparts.

Lacking the rich dower of warm sunlight and magnificent old marble palaces that Venice enjoys, these northern Venetians certainly possess, nevertheless, an exceedingly fine feeling for color, and sense of "values" in painting. Many of them carry their tastes and sense of values into an appreciation of teas, coffee, diamonds and other Asiatic produce just as well. And if you doubt that their painters do possess this appreciation of "values," just price some of their works! You will be apt to conclude that the values appreciated just before you entered the studio.

—One characteristic thing that the old Bible House afforded us, was some real old-fashioned Dutch dishes in its dining. It was a relief to get at last beyond the bounds of the French table-d'hôte menu. Two things still linger in these Low Countries which are surely disappearing from Europe,—national costumes and cookery. Everything in dining and dress is fast becoming modeled uniformly after the Parisian fashion.

It is true that in Switzerland they sometimes dress their waiter-girls in costume, as at the Schweitzerhof at Schaffhausen, but that is sure-enough "dress parade," with nothing more real about it than an opera-chorus of peasant girls in variegated bodices and striped stockings. The Lounger saw "William Tell" represented on the boards of the Grand Opera in Paris, but neither the scenery nor "figurantes" reminded him greatly of Altorf.

—And also, one really gets little impression of Holland as Hollowland, the land that lies below the sea, by surveying the streets of Amsterdam. It somehow doesn't confirm the old Mitchell's Geography idea of boyhood! But take one of those little steamers at the landing near the "Old Bible," and go out, through the "draw," and under the railroad-bridge, into the "Y," and toward the waste of waters, of which that is the doorway; out into the atmosphere of fog and mist which seems to surround and envelop the harbor. There, as your little steamer puffs along, the spectral ships, coasting schooners or great East-Indiamen suddenly appear, and as strangely disappear—while shoreward, in the unremote distance, the massive buildings of the low-lying city, with its towers and spires, loom vaguely above the waste, like the first uprising of island reef and palm above the horizon of a world submerged.

THE FLEUR D'OR.

"MR. HATTERSCHEIDT!"—said Mr. Lincoln, soon after his first inauguration, to a Leavenworth applicant for a consulship abroad—"Mr. Hatterscheidt, I understand you want to go and see your Aunt Werp! Well, that's out of the question, but you're a good boy, and I can send you to milk your Ma's Cow!" So Mr. H. took a fair "Hobson's choice"—and went, a political dairyman, to Moscow.

So runs the story:
> "I do not vouch for the truth you see,
> I tell the tale as 'twas told to me."

The Lounger, however, was not to be diverted from his visit to Antwerp—and thereby had the diversion of meeting a lady not unconnected therewith, the buxom hostess of the Fleur D'Or, the widow Collin.

It happened thus.—We had come down from Rotterdam, and arriving at the station, had taken a cab to the first hotel on the list "starred" by Bædeker. Approaching the Hotel St. Antoine through the crowded streets, we were astonished with the unwonted spectacle of the hotel *portier* actually waving away the new arrival, instead of extending the usual welcoming greeting! Seeking then the next "star of the first magnitude,"—Hotel de l'Europe,— its major-domo appeared at our carriage door, and politely regretted that they, also, were "too full"—not for "utterance," but for the reception—of guests. This was something most unusual, abroad. The explanation afforded was that the town was full, as we could see for ourselves. What was the cause? Was "the circus in town"—or was this "a Fourth of July" celebration, albeit Sunday and the 14th of August? The latter guess wasn't so far out of the way, after all. This was a gala day, a "féte" day for Antwerp. It was their annual civic holiday, of which, it appears, each Flemish town is apt to have its distinct, specific own. Just what historic, memorial significance this day was to Antwerp, we failed to learn, but we did comprehend that it was high time for us to be securing quarters somewhere, before the day grew older and the press heavier. Appealing, therefore, to know our best chance for a room, we were told that probably the Fleur D'Or, a little hotel of the second-class, in the Rue des Moines, opposite the Cathedral, might be the likeliest. Though the last on Bædeker's list, among the "unpretending," we sought it as speedily as possible.

Summoned by our driver, the good landlady met us on the curbstone, and then and there, with her arms akimbo, held a parley and negotiation with the Lounger;—if, indeed, that could be called a negotiation which simply dictated terms of surrender unto him. It was a dialogue carried on between good Flemish and bad tourist-French, and in every way the Lounger was at sore disadvantage.

Had she a room? Yes—one, and just one! A good large front room, on the first floor, but—

But what?—The price, Monsieur, is twenty francs! Twenty francs, for inferior lodging, when the best rooms in the best hotels are usually but five to eight!! *"Oui, Monsieur, vingt francs juste."* *"Mais, Madame c'est tres cher, c'est trop cher!"* *"Oui, Monsieur, c'est tres cher, mais,—que voulez-vous?"*—She owned that it was out of all reason, but, *"que voulez-vous,"* "what was the Lounger going to do about it?" The discussion waxed warm. She condescended to explain that she could put in any number of "cots" and "shake-downs" in that room, and thereby realize easily twenty, perhaps even thirty francs thereby, for the town was *"toute pleine."* To the Lounger's remonstrance that the price was exorbitant, she would reply with the *non sequitur*—*"Oui, Monsieur! Mais, que voulez vous,"*—and as she nodded her head when she said it, the Lounger gave in, and she took us in.

Then, later in the day, she found that she had overreached herself, after all, for the town was indeed *"toute pleine,"* and she had applicants by the score, while her great *chambre* was gone! "Monsieur, twenty francs is a great price to pay for a room!" "Truly, Madame Collin, it is a fearful price, as I told you before, *mais—que voulez-vous!* One must sleep somewhere!" "Yes, Monsieur, but I have another room, not so large, but very comfortable, and it is a front room, too. Monsieur can have that for ten francs instead, and then I can accommodate several officers in the large chamber." Finding that the smaller room was as represented, we obligingly made the exchange.

But in the interval, we had sallied out of the crowded tavern into the crowded streets of Antwerp. Everywhere noise and jollity prevailed. A military band was playing all day long in the Place Verte hard by. More bands of music pervaded and paraded the streets, and everybody was tooting or singing. The city populace were out in full force, and the people were "in from the country,"— all in their holiday clothes and preternaturally festive.

Whenever the miltary companies were not marching, the privates were imbibing beer and the officers were drinking wine; and yet, so far, there was little intoxication or disturbance, and at the worst, the intolerable "fireworks" nuisance was happily absent. On the "*Place*," in front of the cafés, the eating and drinking groups, with their tables, usurped the sidewalk and even projected themselves into the street. Everywhere the Belgian tri-color—black, red and yellow—was flung out from windows and draped across the street, and in flag and streamer, floated gayly from the top of the grand old tower of Notre Dame, which, ever and anon, with its great bells or ringing chimes, adds an old-time peal or merry jingle to the general clangor.

It was somewhat interesting to watch these public festivities, and to note, even here among a foreign people, the different degrees and manner of enjoyment manifested by differing types and individuals. But the noisy demonstrations became rather tiresome at length to a Lounger whose ear never was thoroughly attuned to all the esthetic melodiousness which exists in noise of the noisiest kind, and who has even been rash enough to question, at times, the signal appropriateness of firing off guns on Christmas morning! So, at length, he confined his walks to the more secluded streets, and spent some pleasant hours in the shadowed peace of the Art Museum and the great Cathedral.

Here, in the home of Rubens and of Vandyke, one would naturally expect their best productions to predominate, and this much may be said for that renowned Flemish master, the first named, that here are to be found those of his pictures which evince some quality of nobleness, some depth of religious feeling, some attributes of true greatness in conception and treatment of worthy subjects. After long wearying, in many picture galleries around Europe, of his florid and fleshly Venuses, and aching over his acres of allegoric inanities in the "long gallery" of the Louvre, it is a great relief to come at last upon something that seems to justify in some measure his repute as a great painter. As the curtain rolls down in front of his "Descent from the Cross," hiding it from view, we feel for the first time, a real regret in losing sight of one of his creations. Here too, in Antwerp, are other works of the artist, almost as noble; as well as worthy examples of Vandyke, in whom the Lounger had long taken more delight.

* * * * * * * * *

Perchance it had been the sight of so much flag that sent the Lounger, later in the afternoon, down to the wharves of Antwerp. Here, the "lazy Scheldt" rocked upon its lazily-heaving tide full many a stately craft from other lands, a very forest of masts and smoke-stacks, adorned with strips of bunting, emblems of nationality, conspicuous among which fluttered the Union Jack of Old England! At last, out toward the middle of the stream, the Lounger discovered a single lone specimen of the Stars and Stripes. It covered, on closer inspection, more petroleum, perhaps, than patriotism,—and thereby represented Monopoly quite as well as the Nation; —but it was the first American flag that the Lounger had viewed for many a long day, and he was willing to take it on trust, even if it was an Oil Trust,—and to sing,

"Forever float that 'Standard' sheet."

"How shall we sing the songs of our country in a strange land," when the only two commercial representatives you meet abroad are a barrel of coal oil and a Singer sewing-machine? Never mind, we will be patriotic, all the same!

"Though we may forget the Singer,
We will not forget the song."

Returning from the quay and its net-work of curved streets, the Lounger purposely took the wrong turning, losing his way, and abandoning himself to every passing whim, being sure to encounter, at every step, some quaint and curious building, down here in the old quarter whose suburb had haply survived all the perils of two great sieges, and the fiery ravages of the memorable "Spanish Fury." Grim and gaunt-looking indeed, are these old warehouses, with their stone casements, small, square windows and gable ends to the street, rising with pyramidal steps to a peak far overhead! And when one reaches the "Grand Place," here are more which add a picturesque mediæval beauty to quaintness and grimness; the ancient Guild Houses with varied architecture, in the vicinity of the extraordinary old Hotel de Ville, and one with beautiful symmetric facade and a peak piercing skyward, which they tell you is of the fifteenth century, and name as the Palace of Emperor Charles the Fifth.

But, go where'er he will among these crooked streets and strange surroundings, the Lounger may never quite lose himself,—for everywhere a shining land-mark, away above the roof-tops and pinnacles of buildings, soars heavenward the glorious tower of Antwerp Cathedral! And were it even lost to sight, you are inevitably and pleasantly reminded of its vicinity, for ever and anon, it sends forth its happy chime, and at longer intervals, one of its great bells peals out in deep bass, magnificently sonorous, while every passing fifteen minutes is measured off by three rhythmic cadences, repeated for each quarter: Ding—Dong—Bell!

As we near the entrance to Rue des Moines (the "Street of the Monks") and our inn of the "Golden Flower," we are tempted by a little by-street that curves to the rear, and which was probably the wheat market in ancient days, for it still bears the cognomen of "Vieux Marche au Blé." Here is haply a little eddy in the rushing, brawling stream of "fête-day" festivity, and we gratefully seek its quiet. Here the clamor is hushed, and something like a Sabbath stillness seems to prevail. Neither is this detracted from greatly, after all, by our coming upon a little group of "children of the quarter," who have taken possession of a portion of the narrow street, and are playing some childish game which is new to us, but may be centuries old in Europe. They have set some lighted candles upon the ground before them, for a center, and then, joining hands, are circling around, dancing and chanting merrily. They are only foreign "street gamins," indeed, but their evident innocent enjoyment, at this twilight hour, was the "one touch of nature" which somehow reminded " Queen Consort" of one little boy of like tender years, some thousands of miles away, across the broad Atlantic!

Afterward, we sat, in the gloaming, at the window of our chamber in the Fleur D'Or, and looked across at the little booths which cluster against the outer wall of the old Cathedral, for all the world stuck thereon like mud-wasps' nests. And the crowds yet thronged past, in the street below, still keeping up their holiday. The indications were that some, not satisfied with a day of it, were bent on "making a night of it" also! As dusk settled down upon the streets of old Antwerp, the pace of their fun grew faster and more furious. Their spirits were evidently rising as the wine and beer and spirits went down. Longfellow tells us in sweetly measured numbers, of what, betwixt sleeping and waking, he haply heard and haply dreamed, of the Old Belfry of Bruges:

> "As he lay
> One evening at the Fleur de Ble."

But the Lounger will scarce attempt to put into rhyme what went on that night around the Fleur D'Or of Antwerp! It was altogether too festive for poetry, or for sleep to tired eyelids. Probably on other occasions this little Flemish tavern was orderly enough, but to-night it was "just howling," inside and out! The racket was kept up on the streets till three in the morning, while almost up to that late hour, the officers who had taken the great "*chambre*" just across the corridor from ours would come trooping up, drunk and noisy. At intervals too, far into the night, there would come an awful thundering at the street door, as clamorous as that which, together with guilty consciences, once terrified Thane and Thaness Macbeth;—and other parties would imperatively demand shelter for the night, not to be denied until Madame herself descended to assure them that her inn was indeed "*toute pleine.*"

And ever and anon the waning hours, with our waning chance for slumber, would be tolled off to us by the great clock in the Cathedral tower hard by,—at times by great, deep-toned Carolus himself. And constantly, the quarters would be intermitted to us by the cadence of the lesser bells;—Ding—Dong—Bell! Ding—Dong—Bell!

LAST INNINGS.

STRAIGHT as the flight of an arrow or an eagle launched from the crags of Holyhead Mountain, our little steamer goes skimming the light billows of the Irish Sea, never turning or doubling in her course, though always on the way to Dublin.

Returning to Liverpool after some months of wandering about Europe, we had left there, at the Cunard office, our baggage and other plunder of the campaign, to be put on board the steamer sailing a few days later, and had set out upon a little run into Wales and Ireland, to meet the vessel when she should touch at Queenstown. By Chester and the "Sands of Dee;" by the bold Welsh coast, with its picturesque old towers and ruins; by the vale of Llanberis and the foot of Snowdon; by historic Carnaervon Castle, and the Victoria Tubular Bridge, over the Straits of Menai;—these were pleasant stages which had brought us to the marshy islet of Anglesea, and to barren, rocky Holyhead!

Arrived at Dublin, by way of the little railroad from Kingston, we find quarters at "The Gresham." Like the traditional "fine old Irish gentleman," this hostelry may be characterized as highly respectable, but, compared with its English congeners, somewhat free and easy in its ways and manner of keeping. We cannot say that we liked it any the less therefor, finding in it, indeed, some suggestion of American hotels long left behind, but soon to be regained. And one thing it served at its table, which in turn served as a suggestion of home—good soft bread. We were reminded that this was a blessing which had "brightened as it took its flight," coincident with our taking a Cunard steamer at New York. Like other crude Americans "fresh" to European ways, we had vainly hankered after salted butter and fresh bread, but had perforce been conditioned into something like the converse. It had been a hard trial for our teeth; but like other hard but wholesome experiences of life, we had found it not wholly "stale, flat and unprofitable."

The Gresham is in Upper Sackville Street—and Sackville Street with Phœnix Park make the pride of Dublin. For its three or four blocks of length, with its avenue-like spaciousness, its imposing edifices and monuments, it bears the air of considering itself very fine indeed.

In the early morn, when the "dew was on the gowans lying,"—if haply there were any gowans thereabout,—we drove out to Phœnix Park, which one of Lever's (or was it Thackeray's) characters used to boast as "the foinest in the Three Kingdoms, begorra!" Now, it is forever unhappily associated with the assassination of Lord Frederick Cavendish. Truth to tell, we saw less that was extraordinary in these public pleasure grounds than in the private houses lining the street leading thereto. Behind each semi-circular clear glass of their front-door transoms, shone resplendent a "plaster-of-Paris image," the statuette of some animal—usually a horse or a dog! The esthetic fancy reminded one of Coleridge's pious application:

"He prayeth well who loveth well,
Both man and bird and beast."

*   *   *   *   *   *   *   *   *

Traversing the southern half of Ireland diagonally from Dublin to the mountains of Kerry, near the southwest coast, the Lounger derived an impression which, though cursory indeed, is probably not wholly fallacious, that entirely too much is demanded from the Island! However fertile naturally much of the soil may have been, and favored with an exceptionally equable distribution of moisture, it is yet too evidently a limited country agriculturally, and one wholly inadequate to support a large population. There has to be subtracted a vast deal of waste, irreclaimed or irreclaimable land, and far too much of bog, and barren hill, and mountain. For the rest, there is a painful lack of careful, generous tillage, thrift and improvement. With all its perennial verdure, it is by far too much given over to "blossomed furze unprofitably gay." Something will have to be given back to the soil if it shall continue to be drawn upon perpetually for rent money and potatoes and "poteen."

It follows, therefore, "as the night the day,"—or Farmers' Alliances a condition of fifteen cent corn,—that there is bound to be trouble in Ireland until there shall be a considerable reduction in number, either of the landlords to be supported, or of the tenants who have to support both the landlords and themselves from an insufficient soil, insufficiently reclaimed and improved.

The contrast with its sister island in appearance is sorely to the disadvantage of Ireland, yet it must be remembered that England owes a great deal to reclamation and improvement. The lushest and greenest of grasses now carpet meadows of emerald, once, in the days of the Heptarchy, given over to marshy fen and mucky morass. Drainage, underdrainage and persistent high cultivation,—this is the magic that has transformed England from a waste into a garden.

Yet surely one of Nature's most delightful gardens, set in the most picturesque surrounding that a waste can exhibit, is to be found among the mountains and lakes of Killarney! It will not do for the Lounger, at this late hour when he should be covering up the embers of reminiscence, to let loose upon them, instead, the breath of enthusiasm, for they might then chance to glow too brightly for him to escape from their fascination. Far less will it answer for him to attempt prose description of scenes which Ireland's poet, Thomas Moore, long since embalmed in verse.

It was in an Irish jaunting-car, the national vehicle, that the Lounger pair set out gaily in the morning from the Railway Inn, at Killarney village, to "do" what might fairly be "done" of all these wonders within the compass of an autumn day,—one of those days which, in Green Erin, begin with fairly-fair weather, progress into poetic mistiness, and end at last in prosaic "drizzle-drozzle." The fair and misty weather carried us the round of the Lakes, and the drizzle-drozzle found us toiling up the long stretch of the Purple Mountain beyond, from whose acclivity we could look across to the reeking MacGillicuddy's Reeks, and down into the Gap of Dunloe. Two forlorn travelers, lacking umbrellas and a prophetic sense of Irish weather,—one in each pocket of that pair of saddle-bags which is called an Irish jaunting-car,—found themselves perfectly willing, at length, to abandon that mountain, whose chilling mists had pretty thoroughly saturated them. Coasting down then between lake and mountain, the driver brought up at a point in the road opposite the grounds of Hon. Mr. Herbert, M. P., and suggested to the unhappy tourists that of course they wished to alight here and view the

Torc Cascade! Truth to tell, they would have been willing to forego any further sight-seeing—but the tourist conscience cried that here was one more duty to Nature unfulfilled; so they complied, and were directed to a path up the side of Torc Mountain, through its leafy, dripping woods.

Arrived in front of the Cascade, they agreed that, like many other conventional things to be "done" in Europe, it hardly "came up to the bills,"—but congratulated themselves that, at all events, this was actually the very last one! Incidentally some things might be noted by the wayside between this and the steamer, but the turning point had been reached. Henceforth the pony's head would be turned toward Queenstown, and their steps toward Home!

Coming down again to the road and their equipage, and lo! two figures rose up before them,—two gaunt, weird figures, draped in long red cloaks and looming in the mist! "So withered and so wild in their attire," it would be no great wonder indeed, if to the imagination of Queen Consort they assumed the significance of Two Fates standing in the way between her and a happy home, soon to be sought across the sea. So interposed to Macbeth and Banquo, hasting home, fired with success, the Three Weird Sisters on the blasted heath! These sisters were but two indeed, two ragged crones, redolent of whiskey—but they were evidently in the mood to predict either good or ill, according as themselves should be sped. And though the road was a fairly good highway, they were prepared to "blast it" fast enough, if occasion should serve. As to the "witches' brew"—but let us not anticipate!

Now, sooth to say, there be happily one element lacking in the mental constitution of the Lady whom the Lounger knows as Queen Consort. She believes in no foolish superstitions! Those idle, silly fancies of the race, coming out from

> "Desolate wind-swept space,
> From Twilight land, from No-Man's Land,"—

these trouble not her a whit! She wears opals in preference to any other jewelry. To her, it be perfectly indifferent over which shoulder the new moon doth first appear, and whether black cat, or rabbit haply cross her path. Of all things she thoroughly enjoys a comfortable dinner, when just a round "baker's-dozen" sit down thereto. And when she goes on a journey, she always sets out on Friday, wind and weather and providence permitting!

But somehow she seemed a trifle uneasy, and anxious to get rid of those beldams. And so was the Lounger, who favored, not buying them off, but driving off from them, "instanter." As we resumed our seats in the car, they attacked the Lady with the most flattering allusions and adjurations to the brightness of her eyes, and the goodness of her heart, which would impel her to induce "his honor" to take a drink of milk from the jug they produced. Oh yes! do try some of the milk and encourage the poor old women! No! the Lounger didn't care for milk, neither did milk "like" him. Beside—he didn't know the cows! Oh, the co'os are all roight yer honor, an' shure they're goats! Well! Queen Consort might "try" some, and welcome, but it was just that kind of a cold day when the milk would "get left" with him! No, the lady didn't seem to favor the beverage either, but it was just a real shame for the Lounger not to buy some, and help those kind old ladies!

Just then, one of them had gone to a crevice in the stone wall hard by—and produced therefrom a suspicious-looking bottle. Well, then, his honor would shurely thry a glass of Mountain-Dew. It was the raal ould Oirish stuff, yez know, and had niver seen a gauger at all, at all! It would war-rm up his har-rt this cowld day. The poteen would put new loife in him! Oh yes! said the Lady, do try some of their mountain-dew! You do need it after the ride on the cold mountain!—Why, Queen Consort! this is really astounding! A representative of a Prohibition State abroad, to be asked to forego his principles, just for the sake of these old crones, and indulge in mountain-dew, which is but another name for Irish whisky of the smokiest and most illicit kind, for it is plainly illicit whisky they are peddling, and of the most contraband variety. Never, no never! The driver corroborated this impression, and added that if the Honorable Mr. Herbert knew what they do be selling, they would be in Killarney jail insoide of foive hours. Oh! that would be a shame, you know, said the Lady, and really you do need it as a medicine, this cold day! Would she try some of the medicine herself? No, she couldn't think of such a thing—but the poor creatures must be helped some way!

Just here a compromise suggested itself to the brain of the Lounger. Our driver had never been in a Prohibition State or in a state of prohibition, though unusually temperate. He had kindly given over his frieze top-coat for the protection of the Lady. He was the one who needed refreshment, and who took it finally, in a judicious mixture of the milk and mountain-dew;—the chill of the one duly qualifying the fire of the other. And the Lounger's shilling went to the "kind old ladies," who thereupon set up and sent up a loud chorus of blessings upon him and his lady, and their children to the remotest generation. Bliss your swate face, lady—and may the blissed Vargin give you a safe passage across the Ocean, and take you safe back home to the dear childher in Ameriky! All the Saints in the Calendar were invoked to add force to this kindly wish. As the pony trotted off toward Killarney, until a turn in the road hid them from sight, we could still see and hear them, from the middle of the highway imploring these kindly blessings upon our devoted heads.

And, just think now, that if you hadn't bought their "mountain-dew," and given them the shilling, it would be just the other thing they would be sending after us— and we, in two days more, to be out on the wild Atlantic!—said Queen Consort. Oh, but she was deep, —and not the least bit in the world superstitious!

* * * * * * * * *

Next evening found us at Queenstown, and once more stopping at a "Queen's Hotel," the last of our Inns, and the very last of that series which is so numerous in the United Kingdoms as to suggest the popularity of some of those "Queens of England," whose list—and wardrobes—Agnes Strickland hath inventoried.

The next morning the Lounger dropped into the Cunard office, and inquired for any news of the steamer, and of the state of the weather seaward. For the first, the boat would be standing into the roadstead in an hour or two; and there was a cable from the New York Herald Weather-Bureau that a storm wave was sweeping across the Atlantic! But you never find the New York Herald reliable?—queried the Lounger. Well, their weather predictions are generally fairly verified,—was answered, for the Lounger's satisfaction. And this was to be all he was to receive from the shilling that the Lady had made him invest!

Two hours later, and as they looked out into the harbor, there was the Cunarder quietly at anchor, waiting for the Lounger and the last London mail; and just before they got aboard the little tender that took them off, he wrote, in the register of the Queen's, which previous travelers had converted into something of a memorial album, these lines of Good-Bye:—

> Old World! we leave you now with some regret;
> Glad have we been, amid your scenes to roam;
> We own their charms;—but fairer, dearer yet,
> Are the bright skies and spreading vales of Home!

THE CITY OF BY-GONE YEARS.

There is no superstition so wide-spread in Europe as that of a sunken city which has disappeared below the surface of a sea or a lake at some unknown period in the past. When the waters are rough, the tips of the spires of its churches may be seen in the trough of the waves; on calm days one hears the distant sound of their bells, drowned by the ocean. The name of this city in Germany is given as Vineta, and it lies in the vicinity of the island of Rugen. E. Werner has a novel entitled "Vineta," which is based on this superstition. In Brittany this sunken city is called Is, and various places along the coast are pointed out as its site. Ernest Renan has made use of the old legend in his "Souvenirs," as follows: "It seems to me that I have in my own heart a town of Is, which still has its obstinate bells that ring for the sacred offices and call for men who hear no more." —*American Notes and Queries.*

There is an old, old legend
    That beareth a charm to me;—
A fanciful tale that lingers
    By the shores of the German Sea.

It tells of a sunken city
    Whose towers the waters lave—
The grandeur of by-gone ages
    Now resting beneath the wave.

Down many a hundred fathom
    It slumbers quiet and long,
With all its wonders and treasures;—
    A city of story and song!

Its every marvellous palace
    The water-gods hold in fee;
Its halls are the homes of the mermaids,
    Its pavement the floor of the sea!

Oh ne'er to its topmost turret
    Was deepest plummet sent;—
Nor has foot of the boldest diver
    Trod its loftiest battlement!

But sometimes at eve, in autumn,
    When the sun sinks slowly down,
Darting the shafts of his splendor
    On the grave of the buried town,—

When, over the mirror of waters
    Burnished with crimson and gold,
Float the streamers of cloudland glory
    By the legions of Sunset unroll'd—

Then the voices of restless Ocean,
    Sounding from year to year—
Are still'd with the tumult that bears them
    —And a music falls on the ear!

The bells of unseen steeples
    Swing magically to and fro,
Ringing tones of silvery sweetness
    Up from the depths below.

And then, to enchanted senses,
  Through a golden mist, uprears
A pageant of marvellous beauty—
  The City of By-Gone Years!

Pinnacle, dome and belfry,
  Palace and knightly hall,
Fortress with rampart and bastion
  And pennon on castle wall—

The myriad roofs of its mansions,
  Street, column and portal proud—
Float upward, above the sunset
  And hang in the sunset cloud!

Then—wonderful picture of beauty!
  Roof, rampart, tower and spire,
With more than their old-time splendor,
  Glow bright in the sunset's fire!

Awhile burns the magical city
  By the parting sunbeams kissed,
With flushes of rose and of crimson,
  Of ruby and amethyst!

But e'en as we gaze—lo! it fadeth—
  Fadeth the purple and gold,
Swift changing to evening shadows!—
  And the night comes, dark and cold!
      *   *   *   *   *   *

And thus in the Autumn of Lifetime—
By the shore where its sunsets glow,
Through the crystalline waves of Remembrance,
Rise the visions of Long Ago!

'Tis the one, of all teeming fancies,
   That fondliest reappears,
Which Time had buried the deepest
   In the grave of our long-lost years.

And sweetest of mortal music,
   With tone that clearest swells—
The chimes of our happy childhood,
   Rung upwards by Memory's bells!

How oft to enchanted senses
   Through the golden mist, uprears
That pageant of marvellous beauty,
   The city of by-gone years!

With more than its youthful glamour,
   With splendors grown manifold,
Shine its roofs and its spires, in the glory
   Of fancy's sunset gold!

The steps of its rocky castle,
   The stones of its rugged street,
That wearied each untrained muscle,
   And wounded our youthful feet,

Now shine in the magic radiance
  With glisten of precious stone,
In semblance of marble and jasper,
  Of agate and chalcedone!

Forgotten youth's toil and sorrow,
  Banished its care and pain;
While, brightened with ten-fold luster,
  Its triumphs and joys remain.

But soon, like the somber nightfall,
  Chill age upon manhood crowds,
And memory's Fata Morgana
  Shall fade, like the sunset clouds!

* * * * *

Where then, is the City Eternal,
  "By saint and by prophet foretold,"
Shining aye, with a glory supernal
  Transcending the ruby and gold?

No vision of human fancy,
  No city of earth or of air,
May hint but the faintest promise
  Of marvels the future shall bear!

Each splendor of cloudland shining,
  Each pearl, and each tinted shell,
Prefigures an infinite beauty
  And glory ineffable!

How bright the beatified mansions,
   Where "the pure in heart" take their abode
In that radiant City Celestial,
   "Whose builder and maker is God!"

What language of men or of angels
   Shall tell what its glories may be,
Whose domes arch the Universe endless—
   Whose foundations—Eternity!

"Where is neither beginning nor ending"
   Lo! the spire of its Temple uprears,
Whose chimes, rung at dawn of Creation,
   Commingled the Music of Spheres!

That we, its glad belfry and portal,
   With "the eye of the spirit," may see;—
May dwell in The City Immortal—
   God grant it to you and to me!

IN MIDSUMMER MOOD.

Those things do best please me
That befal preposterously.
　　　　　—*Midsummer-Night's Dream.*

HOW JOAQUIN WALKED OUT.

He was a brick: so said they all,—
And hefty as a brick let fall,
Down dropping through the summer sky
Upon the head of passer-by,
Down dropping through the filmy air,
On brick-top head of passen-jare.

I loved a maid of the wild Tagfasters,—
Bronze-hued, brown-eyed, with lips like wine,
With a soul tip-toed and stretching higher,
(Reaching up to my soul's desire,
Fiercely fond and full of fire)—
And a flat-foot fit for a number-nine.

Let friends be false—let friends be true,
Let wine be old—let love be new,
Let fields be green—let fields be bare,
And sun-shafts shoot through shimmering air,—
With bright skies arching and bending over,
Trill of tree-frog and lay of lover,
The hum of bees and sweet scent of clover,—
Home, friends and all, by the white sea-wall,
To win new loves, or to serve new masters,
I left the Sierras and wild Tagfasters!

MIDSUMMER MADNESS.

THE HEIGHT—and the heat—of midsummer is here! "'The Dog-Star rages." So doth also "ye managing editor," because, forsooth, the Lounger has lately chosen to lounge in cool and shady coverts rather than in "the keen sunlight of publicity"—the columns of the JOURNAL. "Why doth the heathen rage and imagine a vain thing?" Undoubtedly the JOURNAL readers would prefer to grant the Lounger a summer vacation, once for all, to having him parade in these columns in coatless and cravatless mental dishabille of hot weather.

Just why the Dog-Star should rage and *par consequence* the world grow mad, either in July or August, the Lounger has never been able to determine. To him, it has appeared that Sol rather than Sirius was accountable for this torrid heat that fries men's brains and coagulates their wits. Sirius-ly, the Lounger takes no stock in this Dog-Star theory! The hapless victims of lunacy (mental moonshine — *derivation*, Luna — See?) are prone to imagine themselves the only sane, and all the rest of mankind demented. Possibly that is the matter with the Lounger now,—but he does appear to strike a good many of late, afflicted with midsummer madness. Either several people in the world of literature are off their balance, or else the Lounger is slightly off—his base." "It is a mad world, my masters!"

First, here comes Estes & Lauriat, book publishers—

with an advertisement in the "Atlantic." "Do you always know just what to do? If not, let us recommend Mrs. Florence Howe Hall's 'Social Customs,' (price $2.00) and its baby relative, 'The Correct Thing,' (price 75 cents); for, with both these books, one can make no mistakes in life, as every possible question may be answered from their combined wisdom." This is midsummer-madness with a vengeance! It is as astounding as the declaration of the escaped poor-house "luny" who assures you he is King Solomon, on his way to return the visit of the Queen of Sheba, and anxious to sell you one of his cast-off crowns of gold for two-bits, towards paying his railway fare! Verily E. & L. are as luny as a loon; as mad as a March hare; as crazy as a— *Cimex lectularius!*

"Do you always know just what to do?" Just reverse that, please, and put it—Do you *ever* know just what to do? The Lounger has imagined sometimes that he did, but generally discovered his error before long, when too late! As a rule, instead of there being just "a right way and a wrong way" to choose from,—which is easy enough,—it has appeared to him a very perplexing question as to the better of two right ways, or the lesser of two or more evils presented. And then, whichever course he did decide upon usually led him to regret that he hadn't taken another!

Fortified with these two books—aggregating a cost of only $2.75—"one can make no mistake in life, as every possible question may be answered from their combined wisdom." Indeed, this is very tempting—infallibility for $2.75! And yet, Messrs. E. & L., we will not invest! Unlike Mr. Blaine and Henry Clay, the Lounger would rather be President sometimes than to be always right.

The Lounger's next candidate for a Midsummer Asylum is a Reverend Somebody, who furnished a paper for the recent Chautauqua Assembly of Missouri. By the way—is there any other association than that of sound between "chalk-talks" and "Chautauquas?" Possibly only this—that both alike press close up on either hand, to the frozen summit of intellectuality.

The theme of the Reverend's paper was the "Women of Shakespeare," and he had undoubtedly made up his mind to strike it at once, like a cyclone, tearing the whole subject up by the roots and bearing it aloft on the wings of the tempest—and of a vivid imagination. He "struck twelve the first time," and very promptly! "My theme is the Women of Shakespeare. 'Women' and 'Shakespeare' are the two best words in the English language!" Well!—The Lounger will prudently file no exception to the first, in this connection, but in all common sense— and in all reverence, as well—what becomes of such words as God, life, immortality, father, brother, home, friend, faith, love, truth, virtue, justice, honor, and a hundred others that should rise from heart to lip, ages before any personal name, however great, should be spoken or recalled! This is the veriest midsummer-madness of Shakespeariolatry that the reverend gentleman is infected with—and his only excuse may be that other writers who should know better have taken on this silly habit and fashion of loose, exaggerated speech whenever the name of Shakespeare is mentioned. Even so great a man as Emerson himself once fell into it, and wrote some such nonsense.

Our Chautauqua hyperbolist then went on to enunciate his proper pet theory—which was that while the *men* of Shakespeare were often quite faulty in conception, the *women* of the dramatist were always unique and perfect in idealization—with the possible exception of Ophelia—who, in point of fact, went about as crazy as the reverend gentleman himself. On the other hand, some other Shakespearean crank would promulgate that the women are "as flat as dishwater," while his men are admirable characterizations throughout! This is the amusing yet saving point of average Shakespearean criticism;—the commentators, like Kilkenny cats, devour each other. Each admires, and each discards;—what is the poison of one is the meat of another, until, like the good Mussulman's swine—

"Quite from tail to snout 'tis eaten."

* * * * * * * *

There was a man of the name of Stevenson who startled the world not long ago, with his rocket-like ascent into the literary firmament. What has become of the author of that "Strange Case of Dr. Jekyll and Mr. Hyde," "The New Arabian Nights," and "Treasure Island?" Can it be the same hand that once "drew the long-bow" of story-telling with such marvellous power and skill, that now sends such a puerile shot as "The Black Arrow?" This is no tightly-strung cord indeed, whose rebound shall plant the feathered shaft in the bull's-eye center, quivering! On the contrary, it is the loosest sort of string, scarce attached at either end;—and it goes trailing along—one feeble and frayed fold of invention following another with no logical continuity or sequence;—a weak thread spun out weekly. Poor Louis Stevenson! His is no violent case of mental aberration! Failing physical energies find a reflex in failing mental power. He needs "rest and a change."

* * * * * * * *

But the Haggard novel-writing fiend is as gaunt and grim, and as blood-thirsty as ever. His ideal hero reverses the old adage, "it is better to be the first at a feast than the last at a fray," for he is never so happy,—be it Allen Quatermain, Umslopgaas, or who he may—as when, standing in "the imminent deadly breach," he slaughters the natives by the score or by the hundred. "Saul has slain his thousands"—but Rider Haggard has slain his ten-thousands—by proxy, and with "the jawbone"—no, with the pen—of an ass-assin! This is a "bloody Englishman" sure enough! He fairly gloats over murder and carnage. With some illusion of picturesque stage-setting, amid strange scenes, and with all the heightening of a "stunning" style, he idealizes the work of the typical cattle-shooter and champion pig-sticker of the Kansas City packing-house; only, his hero "gets in his work"—and a big day's work at that—on human beings, or at the very least on terrific lions or monstrous elephants. Three of the latter, his latest "big chief" slays in one night, "just for divarsion's sake, me boy,"—to fill in a wakeful hour or two, when it wasn't a very good night for sleeping—or for elephants either!

Whether this apotheosis of brutal butchery exerts any higher moral influence over the mind of youth than does the deprecated "yellow-covered" dime-novel may be a question. What is the essential difference between this and the "Big Bulldog of the Brazos," the "Red Slayer of Socorro," or "King Richard the Third." Their heroes, be they bandit, bully Briton or king are all alike in their appetite for blood,—and their easy royal manner of butchery. "Fee—Fi—Fo—Fum"—"Off with his head!—so much for Buckingham!" As the Lounger has just said of Haggard—there is a good deal of vivid picturing and phrasing,—a rush and a "go" of directness about the style that must be quite attractive to almost any boy:—but the brutishness is there all the same.

For the Lounger, its quality—and especially its quantity—begins to pall. Even the excitement of the terrible conflicts and dangers wears off, when you find that after sacrificing all the poor fellows, that he has driven in to help him, the hero himself always comes out all right in the end.

Long years agone, the Lounger suffered a succession of nightmares. Every night, in his dreams, he found himself frantically clutching, with hands and teeth, the lower edge of the roof of an exceedingly tall house, while his body dangled over the abyss below. It was at first decidedly unpleasant. But he continued to dream this same thing so often that at length a continuity was established, bridging over the intervals between;—that is, he came to recollect in his dream of having been through the same thing before! With this recollection, came even this logic,—"I wasn't killed the last time, or I shouldn't be here now. I think I'll let myself drop!" He did so, and fell—awake! Thereafter, that particular nightmare had no terrors for him—and forsook him.

So with the terrors of the Rider Haggard school of romantic fiction, (a "rider haggard," by the way, is somewhat suggestive of "night-mare") too much familiarity with unlimited bravery, butchery and bugaboos generally, breeds contempt in mature minds.

Possibly apprehending this philosophic truth, Haggard now "gives us a rest"—and a change. In his latest craze, "Mr. Meeson's Will," he treats us to a unique variation on the theme in fiction, of shipwreck and subsequent sojourn on a "desert island." The Lounger had supposed this theme and its incidents exhausted by Defoe, Charles Reade and Frank Stockton; but Haggard discounts "Crusoe," "Foul Play," and "The Dusantes." By one bold stroke he obliterates the "tracks in the sand," breaks Penfold's pearls, and demolishes even the "Ginger Jar."

The mean, miserly Meeson—a British publisher—together with the heroine-novelist Miss Augusta Smithers, a boy, and two sailors, find themselves (about the middle of the book) cast alone on the desolate shores of Kerguelen Land. Meeson, sick and dying, at last repents him of his sins—and especially of his will, which disinherited his nephew Eustace because said young man had chivalrously taken the part of said Augusta, cheated by the rascally publisher out of the proceeds of her novel. He is now most anxious to revoke, and bequeath his two million pounds to Eustace, back in England. Unfortunately it proves there is no paper-mill or stationery-store on this desert island, consequently no material to write the Will. With blood for ink, they might have managed, but there is no substitute for paper,—not a rag of linen among the whole party in flannel—the heroine having left her hem-stitched handkerchief behind, which leaves Meeson in despair! However—"wherever there's a will (to be made) there's a way,"—and Augusta,—inspired by a love of justice and a latent love for Eustace as well,—conceives an original idea. The Will can be tattooed on somebody's back, and thus borne back to England! As no other "Barkis was willin'" to have the Will in, she finally shoulders the responsibility herself, and the sailor Bill—who is an expert in this old-fashioned species of type-writing—tattoos with a fish-bone and cuttle-fish ink a brief but comprehensive will, duly attested, on the ivory shoulders of the fair Augusta. That fair back, resplendent in evening toilet, can never now be presented at court; her radiant shoulders

<blockquote>Once lovely as a statoo
Are now ruined with a tattoo!</blockquote>

This is a new and startling idea in fiction! Let us trust it will never become popular, nor be carried too far, —no matter even if "there is millions in it!"

Meeson dies appositely soon after, Augusta angelically exclaiming: "Well! I'm glad that it is over! Anyway I do hope that I may never be called on to nurse another publisher." The sailors get drunk and drown themselves "out of hand." Augusta is rescued in the nick of time, and finally gets back to England, to bring her fair face— and her back—as a fortune to Eustace. There is naturally a big fight in the courts over this unique Will—but Eustace finally succeeds, of course, being so effectually "backed up" by Augusta Smithers. The contestants had employed some twenty eminent lawyers on their side, who made a great deal of legal noise, but after all, being only lawyers and not drummers, they couldn't beat a tattoo!

The Lounger votes that Rider Haggard be entitled to a ward all to himself,—the First Ward, indeed,—in the asylum for midsummer maniacs.

MIDSUMMER POETRY.

WHEN Matthew Arnold desired to give to one of his propositions the force of an axiom, he was wont to preface it with "I said a long time ago that—." In his mind, the fact that he had propounded it to the world several years before, effectually barred any possible contrary opinion with a statute of limitations; it foreclosed a mortgage upon the world of letters and ideas, and precluded finally any subsequent equity of redemption to any opposing notion whatever.

The egotism of ordinary mortals is apt to be somewhat offensive—but that of Matthew Arnold pertained to so great a mind, and was withal so sublime in its unconscious audacity that we can but wonder and admire!

In humble imitation thereof, the Lounger might say that so long as a year ago, he called the attention of the JOURNAL readers to the fact that literature at this torrid season of the year is apt to be affected with a midsummer madness. The manifestations may be mild in manner and moderate in measure, but—like great wit—they are "to madness close allied." That is, to madness of the midsummer variety, which overcomes the writers "like a summer cloud"—or rather like a July sun when you are abroad without a light cotton umbrella.

Last season, the Lounger gave instances where it touched the brains of the prose romancers. Just now, he is inclined to test some of the midsummer poets.

Some of these are crazy enough in the early spring-time when they sing madrigals to the birds and grass and flowers. Of the typical poet it may be said, as of Bayard Taylor's "Quaker Widow's" deceased husband, "I think he loved the spring." Some of the poets of the last generation were wont to love the distillery even better.

During May and June they have a lucid interval,—but in this month of midsummer, they, break out of their asylums and rove around among the newspapers and magazines. We know how this is ourselves. If the Lounger—like Mr. Wegg—ever does "drop into poetry" it is always in "Boffin's Bower," and at this very season of the year.

Without preface, here goes for the gem of July! It is a piece of sentiment, unaffected, and direct in its expression. Its title is "Lost Light," and the writer strikes its key-note—its *motif*—in the first line:

> "I cannot make her smile come back—
> That sunshine of her face
> That used to make this worn earth seem
> At times so gay a place.
> The same dear eyes look out at me,
> The features are the same;
> But oh! the smile is out of them,
> And I must be to blame."

Now what the Lounger likes about this stanza—and he likes it exceedingly—is its touching simplicity. There is nothing mystical or metaphysical, involuted or evoluted, about its expression. It is poetry of the good old-fashioned sort; you understand just what the writer meant—and it "touches a chord" at once. There's nothing of Browning or the "latter-day poets" about this. So much the better. You can get it into a man's head without a surgical operation, or taking a club—that is, a Browning Club—to him.

Touching, too—though perhaps a trifle tame and prosaic in diction—is the confession embraced in the last line of the stanza. No doubt Edward was to blame! He intimates, it is true, that he doesn't know just why—but if he would consult his own conscience closely, it is probable he could determine the why and the wherefore. A sunshine like that doesn't fade out of a good woman's face without reason! Possibly he had taken to staying out late of nights,—or to chewing tobacco, and has a bad breath in consequence. Possibly—but why speculate? If Edward really wants to know, he can find out.

> "Sometimes I see it still: I went
> With her the other day
> To meet a long-missed friend, and while
> We still were on the way.
> Her confidence in waiting love,
> Brought back for me to see.
> That old-time love-light to her eyes
> That will not shine for me."

Well now, Edward, you see that smile is not "done-gone" for good, and you, too, could share in it if you should deserve it! But you recollect that when you were going down hill, on the way home from the station where you had driven her to meet the "long-missed friend," (not "missed" this time, for the train was on time and made a good connection) the faithful old gray horse stumbled, and then you laid on the lash unmercifully, and swore at him like a trooper. Just then that smile went out of her face "quicker than wink," and fled like "the light that never was on sea or land." Whatever else you let go of, Edward, keep your temper! Hold on to that always! I think you will.

> "They tell me money waits for me:
> They say I might have fame.
> I like these gewgaws quite as well
> As others like those same."

They tell you "money waits for you!" Beware, Edward! 'Tis a set-up job. They will play the confidence-game on you sure, if you're not on the lookout! Money isn't waiting for anybody these days. Most people have to work hard for it; even prize-fighters have to fight for it, and if you invest in Wichita town lots, it is you that will have to wait for the money and not the converse!

"They say you might have fame." Well, you might, and then again you might not. "Doubtful things are mighty uncertain." On the whole, perhaps midsummer poetry gives you just as good a chance as any other way. You confess that you "like those gewgaws quite as well" as does the average person, or, as you tersely put it, "as others like those same." By the way, while "those same" strikes one as reminiscent of Bret Harte's "Truthful James," is it good enough poetry for July even? If this be allowed to stand, the next fellow will be working the phrase "of the which I am which" into poetry!

But the next stanza is tip-top good poetry. It is direct, —it is forcible,—and the sentiment is well conveyed. The Lounger has no words of criticism or censure for the poet when he invokes—

"Come back, dear vanished smile, come back!"

Just bring it back yourself, Edward! it rests with you, for you were "to blame" in the first place! But now for the conclusion:—

"Who wants the earth without its sun?
And what has life for me
That's worth a thought, if at its price
It leaves me robbed of thee."

'Tis a sweet sentiment; but who "wants the earth," anyway? Yet here's a man that wouldn't be satisfied without "the sun," too! Edward, beware of covetousness,—"beware of ambition, by that sin fell the angels!"

The Lounger will scarce go so far as to suggest that if Edward "cannot make her smile come back" he should go away and never come back himself. On the contrary, if Thomas Moore made a success of it by singing "Her bright smile haunts me still," we dont see why Edward's "Lost Light" may not yet come back, and bring "the light of other days around" him—and us. But he must be more careful in every way, and especially with his poetry. He can write good lines, but he shouldn't allow himself to drop into prose and slang—or phrases that sound very much like it. At one moment, with great felicity, he gets hold of words and phrases most expressive of tender pathos—and the very next, with equal facility he "catches on" to "those same." Thus, what might prove a perfect poem, degenerates into an incongruous medley—almost a burlesque. Henceforth let him haply restrain a propensity to punctuate poetic points with the pen of a punning paragrapher!

* * * * * * * * *

For some good sane verse, though born at home and in midsummer, take Mrs. Allerton's "Fields of Kansas." The prairie landscape in all the luxuriance of this—its most luxuriant—season is here well reproduced in word-painting, from a palette well stored with local color. If to the eye and ear of a stranger, there might seem a little too frequent recurrence of the note of "gold" in the landscape of Mrs. Allerton's verse, he would do well to bear in mind that, at even this season of the year, the gold in the sunset sky of Kansas is as common as its rich color is abundant in her harvest fields.

HAND-MADE POETRY.

THERE is an uncommon amount of fairly good poetry on the market now-a-days, considering that we haven't any really great poets "on the hook" in this day and generation,—since that galaxy of great poets that adorned the past era is fast sinking below the horizon. But there is a goodly supply of second-best poetry afloat in the newspapers and magazines, all the while. Not so particularly at this season of the year. The Lounger is not now alluding to poetry of the Midsummer variety, which bubbles and boils and, like the current mercury, comes near bursting out through the top of the thermometer.

No, the Lounger hasn't that especially in mind at this moment! Far less, what is called "machine-poetry." That is not worth considering at all. But what he was referring to was good "hand-made" poetry. There never was a time when more of this kind was produced, and it usually is of a very clever sort, indeed. A large number of the rising young literary fellows of the country are working at it, and the product, on the whole, is very creditable. The only question is—does it supply a long-felt want? Will it fairly take the place of the old-fashioned sort of poetry?

It is indeed, constructed *secundum artem*, by scholars and men of talent, who understand all the rules fully as well as did the old masters of the business. All the various forms are employed with great art. Some forms, indeed, are mastered that the old fellows scarce attempted—the *ballade* and the *rondeau*,—while the sonnet, that highly artistic form of verse, which used to tax the resources of the average poet so high that it exhausted all his income, and nobody but the upper classes consequently had any "outcome" therein,—the sonnet is now-a-days mastered by every young aspiring rhymester.

Holman Hunt records in his recent papers on the Pre-Raphaelite Brotherhood that, early in his literary life, Dante Gabriel Rossetti sent specimens of his poems to Leigh Hunt, requesting his counsel as to whether he should devote himself to literature as a profession. This, as the Lounger understands, is no unusual course with nascent poets. The advice of the gray veteran of literature, given while complimenting Rossetti's efforts, was to the tenor that the profession of poetry was "too pitiable to be chosen in cold blood." Rossetti, like most others in a quandary, sought the most competent advice he could get—and then followed his own inclinations, which in his case led by turns to poetry and to art.

It strikes us, however, that the answer of Hunt intimated an appreciation of the proper function of poetry, and the rationale of its production, a great deal higher and clearer than that of the young inquirer. It would truly be a cold-blooded piece of business to say: Go to, now! I shall make poetry! Poetry by the ream—poetry by the hundred-weight. Poetry shall be the business for me!

The Lounger may be wrong, but it occurs to him that there should be more spontaneity about it than this. The true poet should sing because the song is already in his heart and will burst forth. Of such there is little danger that they will live mute and inglorious, or "die with all their music in them."

Possibly this is but an obsolete view of the art. The

more modern way is for the poet to sit down after breakfast—as Trollope used to spin out a novel—and churn out just so many lines or pages at a sitting, "rain or shine," inspiration fit, or "*non fit.*" And then, take it up again in the afternoon; work it over, and get all the buttermilk out; striking out a sentence here and there (that gave the sense and connection) so as to get the happy effect of elision; turning it over and inside out and working in the "Attic salt" (or the cellar salt) of a paradox or a cryptogram occasionally, so as to render it forceful and rugged like Browning;—or else treat it as a piece of rough-casting and painfully file away at it for hours, and then polish and gild and burnish and otherwise finish it, until it becomes as smooth and decorative as William Morris and Tennyson when in their weak estate.

Rossetti, in his early salad days, we find haunting the British Museum, poring over old romances of chivalry, in order, as he said, to light upon "good stunning words for poetry." One is inevitably reminded by this of Dickens in his boyhood, as little Davie Copperfield, calling at the tavern for a glass of the "Genuine Stunning ale,"—and his subsequent doubt as to whether he actually got the genuine "Stunning" after all!

The poetry that is evolved through much head-scratching, persistent brain-cudgelling, and consulting of rhyming-dictionaries and Roget's Thesaurus answers exceeding well for Carriers' Addresses, but is hardly to be classed as the genuine thing.

Possibly, however, it is not the method after all but the result that is important. There are many roads, with perhaps none of them "royal," into the kingdom of poetry, and the Master, though groping, stumbling or straying wildly in the outset, will "get there all the same."

A BULL IN THE CHINA SHOP.

AN INTERESTING article on the subject of "Revision," in the *Writer*, for April, calls attention, by way of illustration, to the fact that the manuscript of Dickens' Novels is blackened on every page by erasures and interlineations.

The author of said article (it is rather awkward to write of the writer in the *Writer*) has safely skated over the thin ice which covers a deep truth! To change the figure, he has skirted the shores of an unknown continent of truth, of which indeed, he might haply have been the discoverer, save for the mists of tradition and fogs of fraud that veiled it from his vision.

The Lounger, too, has examined the original manuscripts of those wondrous tales, as they still exist in the Foster collection of the South Kensington Museum. He too, noted with surprise, the apparent fact that a world-renowned genius had been under the necessity of revising his words and recasting his sentences, on every page. But the Lounger has gone deeper than this and discovered more. Not only the phrasing in the manuscript has been altered, but—in one conspicuous instance, at least—the thought, the conception of the Novelist has been completely transmogrified! The original one was full of meaning, of beauty, and of strength. This has been malignantly sacrificed and eliminated by erasure, and an entirely new one—weak and meaningless—substituted!

This discovery the Lounger made in the course of his investigations and examinations of the manuscript of that great masterpiece of fiction, "David Copperfield." The reader will recall the case of "Mr. Dick" therein, who is supposed to be the victim of a mild monomania, in which his thoughts, and especially his writings, get inextricably entangled with the gory locks of Charles the First's Head.

This had always seemed to the Lounger a very strange vagary, indeed; one wholly forced and unnatural! Why should Charles the First's head get into the author's head, and through that into Mr. Dick's head? If a *caput-mortuum* was really needed—and a king's head at that—why travel so far back in history as two hundred years? Why not take the first (and last) one that came to hand—the caput of Louis Capet? The thing is absurd—the head is absurd—on the face of it!

Now this "death's head" is brought in to this literary feast as a sort of side dish—an entrée—a pitiable substitute for the original serving of something substantial and sensible! "A Bull in a China Shop," was what was represented as making the real trouble in Mr. Dick's mental warehouse! Here is the passage as it was conceived and written:

"Do you recollect the date," said Mr. Dick, looking earnestly at me and taking up his pen to note it down, "when [that bull got into the china warehouse and did so much mischief?"]

Now, all the latter part of the paragraph—that which the Lounger has bracketed above,—has been carefully stricken out by a pen-mark drawn through it, and the following most weak and impotent conclusion substituted: ["King Charles, the First's head was cut off?"]

Now, this is no mere accident! That notable original text was written with a purpose, and the substitution has been made through design—a deep and dark design.

The Lounger might just as well announce his discovery without further preliminary. The world will have to be startled anyway, sooner or later. It has too long been the innocent victim of a deception, compared with which, Mr. Dick's hallucination was mild indeed! CHARLES DICKENS NEVER WROTE THE DICKENS NOVELS! Their true author was a far greater (and better) man, whose name was but thinly concealed under the pseudonym of "Boz!"

The world is already aware that the earlier sketches were printed over that signature, which however, being to some extent in cipher, was never wholly understood. Charles Dickens got hold of the manuscripts; altered them with his miserable corrections and emendations; published them,—and palmed them off upon a confiding world as his own!

In the meantime, through arts and influences not necessary to be recorded here but which will be fully disclosed in MY BOOK—he had suppressed Boz, and almost entirely silenced him on the subject. Poor Boz wandered about England disconsolately,—never distinctly divulging his wrongs: yet whenever he might hear the impostor lauded as the greatest novelist of the age, he could not forbear exclaiming impulsively and derisively "Oh, the dickens!" (There was a cipher in that). This he repeated so often that at last it became a by-word.

But though he bore the outrage so patiently (for reasons to be unfolded in MY BOOK)—he had resolved that posterity at least, should do him justice and honor, and so took pains to insert in that remarkable passage in "Copperfield," a CRYPTOGRAM—which, properly interpreted, reveals the fact that he, and not the usurper, Charles Dickens, was the author. "The Bull in the China Shop" contains this great cipher! "Boz" had too much good sense to wait and carve this as an epitaph on his own tombstone;—and he was fearful that he might not get the chance to enter Westminster Abbey and place it on that of Charles Dickens. Neither would he take the risk of inserting it in any posthumous edition of the Novels, lest the paging thereof might not prove uniform with the contemporaneous ones. He therefore put it into "Copperfield," as he wrote it!

The Cipher is simplicity itself—when you come to understand it. It does not add up on the "put down a cipher and carry one" principle whenever you strike 10 (this really "strikes twelve the first time");—nor does it progress with first a hop, then a skip, and finally a jump! Like a Limited Express, it starts out promptly on schedule time, stops only at regular coaling and watering stations, and never misses a connection. "It attends strictly to business, and dont go fooling around." It is always loaded and never fouls or hangs fire. There is no other cipher on the market that will do half as much, or begin to do it so well. For instance:—to indicate that it begins right at the beginning (which is the only scientific way of adjusting a cipher—for those that begin in the middle can be ciphered both ways)—B, the first letter in Bull, is the same as B, the first letter of Boz! The rest of this cipher will be found in MY BOOK—1000 pages, all in one volume—sold only by subscription.

Again, "Boz" took a little mild revenge on his base oppressor by satirizing him, in the same place in

A BULL IN THE CHINA SHOP. 277

"Copperfield," as "Mr. Dick." This, you see, is almost Dickens' own name itself! Could anything well be plainer? Under this title he pictures him as maundering over manuscripts and perpetually scribbling on a mass of fatuous nonsense,—which, indeed, Dickens' own writings are in comparison with those of Boz! A great many people to this day complain that they cannot endure the mannerisms of this novelist, overlaying and obscuring as they do, the beauties of the author. Now those are simply the abominable "revisions" and "emendations" that Charles Dickens interlined when he got hold of the manuscripts;—"improvements," he called them, and actually imagined that he was superior as a writer.

No wonder that there was a subtle significance of irony introduced by the real author in this memorable cipher passage! Dickens is compared to a bull who has got into a china-shop—blundering around and ignorantly wrecking and demolishing the delicate products of artistic genius. Moreover,—Boz prophetically foresaw the day when some future American (not Irish) Ignatius, or Lounger should discover the cipher, and thereupon "make a break for" and of, and bring to "everlasting smash" the brittle literary reputation of Charles Dickens. You see there is a double and a treble cipher in this wonderful passage!

Charles Dickens, without interpreting this fully, mistrusted something of it—his dull intellect being stimulated and quickened thereon through the tormentings of a guilty conscience. He therefore erased the passage, and substituted that nonsense about Charles the First's head!

Fortunately, the Lounger has been enabled at this late date, to bring the truth to light;—restoring the Bull and the China Shop to their proper place in cryptographic literature—and the true author to his rightful heritage of fame! "Codlin's the friend—not Short!" (A Cipher!)

This serves to explain many startling incongruities, and reconciles apparent paradoxes that have long perplexed the world. Charles Dickens, the man, has for years been known as the complete antithesis of Dickens the author!

The Writer—the real "Boz"—reveals a genuine man in every fiber of his being:—full of warm affections and kindly sympathies that reach out to embrace all humanity, even in its lowliest conditions.

Charles Dickens, the individual, was a weak and vain creature—a pseudo-exquisite of the first water,—and a good deal of a snob beside.

In the works, the Author stands out nobly distinct as the resolute defender of all the dearest sanctities of home and family. No one in English literature has done more to inspire sympathy for man toward man, or reverence toward woman. The lessons of duty are taught so plainly, that he who reads may run—in the right path always.

Charles Dickens, the neglectful husband, almost broke the heart of a good woman, the mother of his children! Compare and contrast this with the forbearance and devotion so beautifully pictured as exhibited toward a weak and even silly "child-wife" by David Copperfield!

The Novelist exemplified the foibles of an absurdly weak and selfish character in picturing Mrs. Nickleby,— and drew an inimitably funny type in portraying Wilkins Micawber, "always waiting for something to turn up." This was all well enough for "Boz," who was in no wise related to originals of either; but if the reputed man was the author, then Dickens, before the wide world, burlesqued his father and ridiculed his own mother! Charles Dickens had many faults, but he could scarce have been guilty of this! No, no—we must look elsewhere for the authorship.

As well suppose that the deer-stealing poacher of Warwickshire,—the third-rate actor of Blackfriars and the Globe; the vulgar, carousing, money-getting, money-letting, second-best-bed-devising man of Stratford ever wrote the immortal dramas that bear the name of Shakespeare—as to conceive that such a personality as that of Charles Dickens could have existed in the author of the Dickens Novels!

The two men are absolutely irreconcilable. Charles Dickens "knew little Latin and less Greek;"—"Boz," was a "gentleman (which Charles Dickens never was) and a scholar!"

Somebody, indeed, has advanced that "between what a man is, and what he writes, there is no necessary likeness,—no connection of cause and effect." This is evidently but another of those theories, "as absurd as hundreds of other suppositions that are made to fit the" novels to Dickens, and Dickens to the Novels.

There will, of course, be controversy now! The Dickensolaters will not give up without a struggle. The literary world will be divided into a "Boz" party and a "Dickens" party. There will be The-Bull-in-the-china-shop theory of authorship—and The-Charles-the-First's-head theory. But truth is mighty and will at last prevail! It will prevail all the sooner when people buy and read the Lounger's Book—MY BOOK. In that will be disclosed the mystery yet reserved—WHO WAS BOZ?

A FUNNY FRENCHMAN.

THE LOUNGER is endeavoring to maintain some acquaintance with the rising notabilities of the period;—and striving, especially, to "keep up with the procession" of the Great Humorists of To-day.

This is not so very difficult after all for even a slow-going Lounger. The number of the truly great is not large, nor their procession half so long in "passing any given point"—of wit—as the most of them are in arriving at one.

The latest celebrity of this kind is a Frenchman whom parents and god-father christened Paul Blouet, but who calls himself Max O'Rell.

It strikes us at first as something odd that one of that nation should set up as a humorist, though we are free to admit that in wit they have never been lacking. Bright and ready as they are, the French have always proverbially been known to have their Wits about them! Usually, however, they have been quicker to "grasp the situation" than apt to appreciate the humorousness which may lie in one. A Frenchman endeavoring not simply to be funny in himself, but to create fun in and for the world, is certainly an anomaly on the stage of America, as on the American stage; "a spectacle for gods and men,"—and the "gallery-gods" who are wont to view him in conventional aspect as a creature of shrug, grimace and broken pronunciation, should now revise their judgment, and take him, as a good many Americans already do, at his own conception and estimation of himself—for a humorist of the first water.

Max O'Rell has already been characterized by some one as "the French Mark Twain." Given "Every man in his humor," and to his own conception of it, no doubt this would suit Max exactly, whether it would really fit him or no. Palpably he has marked Twain for a model; has modeled himself upon the American humorist as far as possible,—and nothing would delight him more than that "these twain" of different nationalities should be recognized as of "one flesh," or at least, one twin-brotherhood.

If indeed, there were not such an absence of acidity about him, we might approximately term him a sort of Pyro-Gallic Clemens! Imagine Yankee, Munchausen exaggeration superimposed upon French "spirituelle"-ity, and you have, perhaps, the key of the combination and —Max O'Rell!

Without reverting to the first book which gave him repute—a light study of England and the English, entitled "John Bull and his Island"—let us take a glance at his latest, which is supposed to concern ourselves, "Jonathan and his Continent." It is America contemplated by a *flaneur* of the Boulevards—and Americans seen and read through a French monocle.

The very first sentence gives you a taste of his quality and indicates what you may expect in the way of exact statistical information. "The population of America is sixty millions, mostly colonels." This, you see, discounts Sydney Smith and Carlyle for amiability, if not originality, and you note at once that here is one who while losing no time in getting at the kernel (and the colonels) of the whole matter in two lines, is himself, no mere line officer, but a very general at generalization.

However he soon gets down to business and to details, and while some of them are faithful, as obvious, enough, and cutely described, others are, to say the least, astounding! It might seem as if Max, or his collaborator, had filled up the book by borrowing largely the productions of the end-men of American journalism,—accepting all their stories as gospel facts, illustrative of American life, in its true inwardness and entirety. If, indeed, these fellows had laid a veritable trap for him, they could hardly have "stuffed him up" more completely or successfully.

"In vain is the net spread in the sight of any bird,"— but this Gallic fowl walks right into this sort of snare, crowing lustily all the while over his "finds."

The editor of the Critic, though far removed from above class of journalists, caught Max "out" almost before he left his native shores, by requesting an article giving his *preconceived* ideas of America. Max took this in sober earnest as an absurd exemplification of the American anxiety to know, in advance even, the foreigner's idea of our country! Good-naturedly complacent and usually complimentary withal, he is constantly being surprised at these wonderful Americans—and especially at the marvellous Westerners. To him, all those old newspaper "saws" about them and their "wild and wooly" ways are veritably "all wool," though they may be many a "yard wide"—of the truth.

If he had chanced, for example, to read that a Western City Council, going out of town in a body on a junket, had furnished themselves at the expense of their city treasury, with a brass band to give themselves éclat on the way, Max would have accepted the improbable statement as undoubted fact, and exemplified it as the usual and typical thing with American town councils!

There was once a man—his name it was Infelicitas—who deplored mournfully the loss of truth and the decay of faith in the world and in himself. He declared that things had got to such a pass that he sometimes doubted even the absolute truth of what he saw in the newspapers. At least some of it!

He died shortly after. He couldn't live without faith!

There is no trouble of that kind with Max O'Rell. He finds it much easier to believe all the writings of the local editors than to hunt up proof thereof—and he reprints them with great gusto as veritable facts, illustrative of American life.

Not only is Mark Twain his great exemplar, but he is disposed to attribute to that humorist all the venerable publications of Joseph Miller, and all the fossil *Castanea* that have come down or been dug up from the Pre-Silurian ages. All these he credits to Mark—including that characteristic touch about the lawyer, who was found so unrealistic because he had his hands in his *own* pockets! In a foot-note, however, he expresses a lingering doubt as to whether this is wholly new.

New! Why, bless you, Max!—Eleazar, the son of Aaron, told that same thing, in great glee, to a select circle of the Levites, one night when they were all relaxing themselves mightily swapping jokes;—told it to them as a new and especially good one on "the doctors of the law." And, long before that, Abraham had reported it to Melchizedek, as the latest *bon mot* from Egypt—just brought across the desert by caravan. Going that way they had no other loading—and took it as ballast.

But to pass from Max's ancient jests to modern earnest —here is a veritable fact; new, perhaps, to us, though it happened right here, "under our noses,"—though, fortunately not *upon* any of our noses!

"A clergyman in Kansas has just had his nose bitten off by a member of his flock who took exception to some of his remarks in the pulpit." This would be a caution to evangelists—with a vengeance!

Balance this up with an anecdote illustrating the crudity and prude-ity of the effete East:

"A Philadelphia lady, in whose house a gentleman was seated one day at a table, grew red to her ears at his asking her which part of the chicken she preferred, the wing or the leg."

Back to "our muttons" again, the wooly Westerners! Note how tamely we defer to the ladies in traveling:

"In the States of Kansas, Colorado and others, a woman on entering a car will touch a man on the shoulder and say to him almost politely: "I like that seat—you take another!"

Contrast this state of society, and man reduced to the abject submission of high civilization,—with the but recent condition of New York city, where "good-looking young women of the best society cleared its streets from disreputable characters by going out themselves at night unattended, and then striking every man who accosted them."

That was the way it was done! A slight touch on the shoulder is sufficient in Kansas—but in New York city, it required a vigorous striking out from the shoulder to accomplish "what the authorities dared not undertake," so that is now perfectly safe for respectable ladies to go about the streets at night without escort!

Max had this on the best authority—"a lady who enjoyed that most esteemed of woman's rights, the right to be pretty,"—and who gave him "some very curious details on the subject of New York life." Well!—the Lounger should think so—and how much amused she must have been to see him jotting it all down in his little note book!

Was it the same lady, indeed, who gave him those veracious details as to the mode and manner in which young girls "of the best society" manage to secure wealthy or high-born husbands through bulldozing and black-mailing—and of the peculiar laws that aid them in accomplishing it?

O'Rell was over here long enough—say six months—not only to secure many of these choice stories but also to learn several things about our government and laws which had escaped painstaking Mr. Bryce, close observer and faithful student though he was. But then, "John Bull," even when off "his island," can hardly hope to rival "Johnny Crapaud" of continental scope and intuitive sagacity! For example:—

"Execution by electricity has just been adopted by the governor of New York." Query—is the "governor of New York" an absolute, or a limited monarch?

"During four years, the President has almost *carte blanche*. He can declare war and stop legislation." Clearly our President at all events is an absolute monarch —"during four years!"

Of course Max was very much impressed with the differing characteristics of our great cities—and recites them off-hand very glibly indeed.—"In New York, it is your money that will open all doors to you; in Boston it is your learning; in Philadelphia or Virginia, it is your genealogy." This was striking and original—with the man who first discovered it—but Max weakens its force somewhat by the tautologic corollary—"therefore if you wish to be a success, parade your dollars in New York, your talents in Boston, your ancestors in Philadelphia and Richmond." Then what will you have left for Kansas City, Max? Possibly you wont need much!

Our traveler admires Boston ideas however, (pronounce "Boast-on," he says,—which may be a French *jeu-de-mot*)—and adds,—"The moral sense of the people will triumph: Boston, not Kansas will win,"—which is bad for Kansas. Maybe Max is a resubmissionist!

Of course he couldn't get out of Chicago without repeating that wonderful story of Mrs. O'Leary's cow which "kicked the bucket" in 1871--"when Chicago had about 100,000 inhabitants."

The Lounger would, perhaps, be creating a wrong impression concerning this book, if he failed to mention that, after all, it is somewhat entertaining and really contains some truth mixed up with its nonsense and fiction. There are certain aspects of American society which are so palpable that "he who runs may read," even though he "runs on" as fast and lightly as Max O'Rell. The be-diamonded-at-breakfast and decolléted-in-afternoon women; the high-pressure, fast-living men; the wash-ladies and the "duchess" servant girls; the train-boy nuisance and the fool sleeping-car porter; even the hotel table-girl of the "second-rate-towns," who brings your indigestible dinner, selected from the magic formula of verbal *menu*—"Troutanturbotshrimpsauce roastbeefturkeycranberrysaucepotatoestomatoesappletart mincepievanillacream;"—all these are set down faithfully, needing no exaggeration. To a foreigner they are strange enough, just as they are!

"All gall is divided into three parts"—as Cæsar told us long ago, before it became so concentrated. This modern Gaul has already shown us his three literary sections—"John Bull and his Island," "John Bull, Jr.," and now, "Jonathan and his Continent." And yet, if not wholly complete, let him quadrate these with a final "Jonathan, Jr.," devoted entirely to our Western States, and to fancy—and not disfigured by any faults of fact whatever!

For this, it will not at all be necessary to make another trip over on the "Germanic." It can all be done by the Collaborator; for this Gallic lion has an American jackall, who figures on the title-page as Jack All-yn,— *he* can compile the whole work from the columns of the "Arizona Kicker," and it will be all right.

A COLLABORATION.

WITH THE mercury mounting madly up to par, and above it; with red-hot meteors plunging and ploughing earthward; with a torrid sun sizzling and frying all the livelong day; with lurid planets burning in the evening sky; with all nature, ourselves included, parched and roasted within a "dry belt," from which every passing rain-cloud flees affrighted; with all this, the Lounger is prudently determined not to take life or literature just now any more seriously than he is perforce obliged. Beyond the range of sun or star,—the milky-way of the midsummer magazine or the nebula of the daily newspaper,—he shall not venture far in his reading, and should he haply encounter an occasional book whizzing through space, he shall endeavor to get out of danger of being crushed by its weight, rather than to venture upon any critical examination of its constituent elements, or to label it scientifically for any literary museum.

One such volume the Lounger came in contact with the other day, and it struck him—on the whole not unfavorably. Just sufficient to leave an impression and yet not heavy enough to hurt. There should have been a mighty magic in it, for it was all about the "Master of the Magicians."

It is a Collaboration—of which there have been several notable instances in modern story-writing. George Sand and Alfred de Musset wrote, for awhile with each other, and then against each other. Charles Dickens and Wilkie

Collins found therein "No Thoroughfare." Charles Reade and Dion Boucicault thereby produced "Foul Play,"—which was no play at all, though afterward dramatized. Then there was the literary "Combination" of Erckmann-Chatrian; of Walter Besant and James Rice; of Mark Twain and Charles Dudley Warner, in the "Gilded Age." These literary partnerships usually last not long. The one that invented Colonel Sellers was good for that trip and Train only. The Erckmann-Chatrian firm fell asunder, for the partners fell out. In our latest and most modern instance, however, the Lounger trusts they may not, for "it's all in the family," and they have just come together. It is Elizabeth Stuart Phelps and Herbert Ward whose life partnership has now resulted in a literary Collaboration. Success to them!

Miss Phelps had long been a popular and, on the whole, a very readable story-writer. Not to speak of her many other —and better—stories, everybody remembers her "Gate" series: The Gates Ajar, The Gates Between, Beyond the Gates. These have sold well, especially the first, which has reached its seventy-fifth thousand! "Born to the purple" of genius, winning both fame and the favor of public and publishers, Miss Phelps has happily tasted some of the sweets of "royalty." She has probably had no reason to be dissatisfied with her share of "the Gate money."

But to return to our Master of the Magicians—"which his name it was" Daniel. This is a story of Babylon. It has been termed an historical romance, but in truth it occupies itself very little with any sequence of historical events, though dealing quite intimately with a certain period of Babylonian history. It is a very vivid picture, almost "a set-piece," indeed, illustrating life in the ancient city. It paints first Nebuchadrezzar at the height of his glory, and then precipitated to his downfall. The scene of the interpretation of his vision; the building of the hanging-gardens for his Median queen, and the final lapse into memorable madness;—all these are picturesquely presented.

The writers disclaim in the outset any attempt to preserve the unities of Biblical chronology. There have been too many school-masters abroad, too many cuneiform inscriptions found and deciphered. They will let the Biblical chronology take care of itself. If this were all, the Lounger might let it pass, but from other things set forth in the story his fears are excited that one or the other of the united authors may not be strictly orthodox. Which is it, the guardian authoress, or her Ward?

Daniel is presented as a pronounced hypnotist, and this serves to account for some of his manifestations. He sways and influences people by "magnetizing" them, and he himself goes off into trances, with or outside of his own volition. It is true the collaborators prudently leave matters in shadowed doubt sometimes as to whether Daniel's trances are really prophetic or hypnotic—with the intimation that, like those of most "seers," there probably existed a "collaboration" of the two. But at all events, Daniel was really one of the Magi of the King's court, the greatest of them all, the Master of the Magicians.

The incident of the "den of lions" is—perhaps, prudently—not introduced. That might have been a little more difficult to treat satisfactorily on the hypnotic theory. Daniel is, however, introduced to the King's zoological garden, where the caged lions are let out singly for the King to pursue and kill. In this chapter, Daniel actually does come into contact with one of these, slays him, and thereby saves the King's life. He is brought into very close quarters with the fierce brute—but, with the writers' skill in description, it is always possible to distinguish "Daniel from the lions."

The fact that there is an actual disease and form of madness, to which a real medical name has been affixed, *Lycanthropy*, in which the sufferer conceives himself transformed into an animal, and simulates the corresponding actions and habits,—this serves fairly to account for the fate of Nebuchadrezzar, "turned out to grass" of his own volition, and without miracle either called in, or called in question.

There have long been so many "burning questions" of theology appertaining to Daniel and the Book of Daniel, which flame up when the torch of investigation is applied, that these collaborators have done wisely in distracting our interest to other romantic characters that had not previously found a place in history. Of course there is a love story, and Daniel the magician fails not to be touched somewhat by the mighty magic of love, over which he triumphs in magnanimity, but emerges not quite "heart-whole, with the least little touch of" regret, at the end of the book.

Without entering into any criticism of the style in which the story is written, the Lounger wishes to submit one paragraph for information. Daniel is being described:

"For so young a man he was eminent as a scholar; quick in acquiring a difficult language and the science of the observation of the stars contained therein."

Now the Lounger being unusually dull, cannot distinguish those stars or determine their location; nor which was "contained" in the "difficult language"—the stars, the observation, or the science! He leaves the passage to his readers. It may strike them so forcibly that they shall be able to see stars "contained therein" quite plainly.

Here is something a great deal clearer,—a literary coincidence! Here is where Bulwer got a striking incident for his play of Richelieu! He unconsciously borrowed it from the Magician Mutusa-illi, in the "Master of the Magicians." Mutusa-illi had been the "head man," the "boss" of all the Babylonian Magicians, before Daniel discounted and deposed him! When failing to discover and interpret Nebuchadrezzar's vision, to cover his disgraceful defeat and to save his life, he drives back the king's guards and minions ordered to seize and execute him and his companions—with this dread announcement: "Here we stand. I draw about this people the mystic circle of the skies. Fall back ye slaves and soldiers! Step but a sandal within this awful ring and Raman blasts ye in his wrath! Backward, I say! or I hurl the curse of Ana, Il and Hea at you, and your first born die!"

Now this antedates Richelieu and launching the famous "Curse of Rome," by several centuries. What new historical romancer next steps forward to put "The pen is mightier than the sword"—say, into the mouth of Jacob, when he had got most all of Laban's flocks and herds corralled so cleverly!

—The Lounger recommends—in good faith—"The Master of the Magicians," as good midsummer reading. While it is sufficiently sensational, its excitements can always be tempered for the hot weather by the reflection that things have had time since to cool down. With the romance of three thousand years agone, you are pretty sure that all the historic characters embraced therein are now departed, and with respect to the others, that if they havn't got through with their troubles yet, it is about time that they were about it.

THE TAMING OF A SHREW.

> "Peter, Peter, pumpkin-eater
> Had a wife, and couldn't keep her;
> He put her in a pumpkin shell—
> And then he kept her very well."

THE WORLD has long lacked due comprehension of the writings of William Shakespeare. This is chiefly the fault of our literary men who have persistently ignored this author and refrained from mention of his works. Were the Lounger's ability commensurate with his zeal, this neglected writer would soon take his true place in literature, and his writings instead of being simply caviare to the general would become the chosen chowder of the high-private—even in the rear rank.

The only trouble with Shakespeare is that he needs enlightened interpretation—and then he is "all right." The Lounger will make a beginning with an essay toward expounding that really charming work—no, play—of his, entitled "The Taming of the Shrew."

The play proper opens with a scene in Padua in which appear nearly all the chief personages—Baptista Minola, his daughter Katharina the Shrew, her sister Bianca, with two of her suitors; also Lucentio and his confidential servant Tranio. 'Twere well now, so we may have a proper comprehension of events soon to follow, that we "make sure of our ground" in the beginning. Much is lost in reading Shakespeare by running too hastily and cursorily over lines that seem light and unimportant! In Shakespeare every sentence and every word is pregnant with meaning. For example, it is stated in the outset that this scene is laid in "a public place in Padua." Now what is meant by a public place in Padua? Let us by all means stop and consult those who have given time, thought, and the fruit of much learning to the careful consideration of these important questions. Dunderkopf, the great German critic, in his invaluable commentaries on this play (6 vols. folio—Leipsic, 1709) states that by a public place in Padua, was undoubtedly meant one of the public squares or open spaces at the intersections of the streets. Kreysigg in his later work, "*Vorslungen Uber Shakespeare*," and Ulrici in his "*Shakesperi Dramatische Kunst*," concur altogether in this view; and as Gervinus, Schlegel and Tieck, as well as the great body of the English and American commentators advance nothing to the contrary, we may consider this question as nearly settled as is possible with any fact alluded to by Shakespeare. Behold then on one of the public squares, or on one of the street corners of Padua, the Minola family and their suitors assembled!

However often the path of Shakespearean research may be traveled and closely scanned—the pains-taking and observing student can always find new and unexpected beauties at every step, and have his eyes dazzled by new illuminations. Not only does Shakespeare "hold the mirror up to nature" of his own age, but deftly turning it in the bright sunlight of genius, he converges the vivid rays, and flashing them far backward across the dark absym of time into the dim recesses of the past, he irradiates every nook and corner with such effulgence that we read as 'twere to-day all their secrets forgotten by history! In this way, manners and customs long obsolete are once more brought to the light of day!

A beautiful illustration of the service Shakespeare renders thus incidentally, is afforded by this very scene. Were it not for its beneficent developments, who would ever have known that it was the custom of the higher classes—the best society of Padua—to rush out pell-mell upon the public squares and street corners whenever they wished to discuss their family disagreements and most private affairs, including their wooings and weddings! Such, indeed, appears at all events to have been the habit of Baptista Minola, "his custom always of an afternoon"—and his family were by no means "backward in coming forward" to follow his frank and laudable example. Baptista Minola, though not wholly a model character possessed certainly one praiseworthy trait,— all his family business at least was transacted "on the square." Possibly there was a well and a town pump in this "public place"—for thither, like Rebecca of old, he brought out all his "family jars." But soft! possibly they came into the cool, fresh air outside, because Katharine had just made the house too hot to hold them! It may have been a "retreat," and not a "reveille."

At all events, this gives strangers, such as Lucentio, a fine opportunity to learn all about the family, and this introduces him easily and naturally into the play. Lucentio, in his dialogue with his servant, discloses that he has just arrived from Pisa to study philosophy in the famed schools of Padua, and his first lesson begins with his learning all about the family affairs of one of the wealthy citizens of the place, but ends with his falling incontinently in love at first sight with the beautiful Bianca. The state of affairs in the Minola family, as developed by their free and easy manner of discussing them, is briefly as follows.—The younger daughter Bianca has two suitors, Gremio and Hortensio, of whom, the former is old and wealthy. Old Baptista, for reasons best known to himself, is determined not to let the younger girl marry anybody until he has first disposed of the elder, Katharina, who, though not ill-looking, has somewhat badly spoiled her matrimonial market by an infirmity of temper that has gained for her the sobriquet of "'The Shrew," and sometimes even that of "Katharine the Curst." Though rather "hard lines" on the suitors, they are compelled to submit to the decree of "the old man," who, in the meantime, is disposed to treat his daughters fairly, by providing them with private tutors in literature, music and mathematics;—from which we infer that the school of Padua, less liberal than those of Bologna, Salerno, and the University of Kansas, excluded women from its walls.

In the second scene, Petruchio, the hero of the play is introduced as just arrived from Verona, and knocking at the door of his friend Hortensio, which gives occasion to a wonderful punning on the phrase "knock me here." Petruchio, who had just succeeded to his father's fortune, has come to Padua to mend it, and being informed of the chance with rich Baptista's eldest but shrewish daughter is more than ready to jump at it. How eager this noble gentleman and model hero is to marry on any terms so they but meet his avaricious greed is fairly intimated in these lines:

> "Be she as foul as was Florentius' love,
> As old as Sibyl, and as curst and shrew'd
> As Socrates' Xantippe—or a worse
> She moves me not, or not removes, at least
> Affection's edge in me—were she as rough
> As are the swelling Adriatic seas!
> I come to wive it wealthily in Padua;
> If wealthily—then happily, in Padua."

This forcibly strikes the key-note of the play. We now readily apprehend thus early, that Shakespeare has as usual, a great moral lesson to enforce; that a woman who dare have a will of her own must be taught subjection to the higher and noble powers; and in this case the hero Petruchio is the worthy instrument of its accomplishment! It is soon known to the suitors of Bianca that Petruchio has undertaken the task of capturing the Shrew, which will then leave an open field for the wooing of her more favored sister, and they joyfully engage to bear the expenses of his courtship.

In the second act, the plot begins to develop. Petruchio offers himself to old Baptista as a suitor for the Shrew. This is almost too good news to be true, and the square-dealing old Italian actually thinks it right to warn the suitor of the peculiar nature of his daughter,— unconscious of the fact that Petruchio is prepared to welcome fire and brimstone, if need be, to win the ducats of dowry that go with Katharine! In the meantime, there has been some ingenious shuffling of parts among the personages of the drama. Lucentio has changed dress and character with his servant Tranio, and now appears as a candidate for the position of tutor of languages to Bianca, while Tranio as Lucentio comes on the scene as another promising suitor for this attractive lady. The nature of Shakespeare's representations of high-toned Italian noblemen is such that a servant has no trouble whatever in adapting himself at a moment's notice to the part, and performs it just as well as his master. As the countryman remarked the next day after seeing, for the first time, Othello performed—"wal, I swow, if they didn't have a darkey in that troupe—but I didn't see but he acted just as well as any of the rest of 'em."

Our old friend Hortensio comes in disguised as Tricio, the fair Bianca's teacher of "music and mathematics."

In the next act, these two gentlemen tutors make the most of their opportunities to woo the gentle Bianca,— in which competition, the noble Lucentio comes out "easily first." However, as the fair Bianca is little more than walking-lady in this great play, we pass her love affairs by, for the present, and turn again to those of her sister, who had previously given Hortensio a touch of her quality by whacking him over the head with the lute he had been instructing her upon. Then there was "a little rift within the lute," and an "interlude" not very full of harmony—until Petruchio commenced his wooing. It is not within our power to do justice to the gentle phraseology of this courtship, wherein such choice endearing terms as "ass," "jade," "buzzard," "turtle," "cocks-comb," and "fool" are lovingly bandied back and forth, until, fairly conquered, or at least silenced at last by such delicate and convincing evidences of his ardent and true love, Katharine finally allows the wedding day to be named! Whilst Katharine is wholly justified in afterward describing to her father and the rest, Petruchio's treatment of her as that of an unmitigated liar and madcap ruffian—yet, inasmuch as Shakespeare makes her give a virtual assent to the proposed marriage with him— we infer, either that our author considered this a model of successful "sparking", or else his Katharine was far less of a shrew than her reputation gave her credit for, and she was really just crazy for a husband. In either case, it is our privilege to watch with unceasing admiration the development of this wonderful idealization of a woman's nature, as portrayed by that great Master, whose province it was to search out the inscrutable mysteries of mind—especially of a woman's mind; and who is never so truly great as when most difficult to comprehend—for in such case how innumerable the springs of action suggested to us by the difficulty of apprehending the proper and peculiar one.

The wedding is set for the following Sunday. Petruchio goes off ostensibly to Venice "after rings and things and fine array." This affair now being in such fine trim, old Baptista is reminded that he had promised in such event to adjust matters between the suitors for his daughter Bianca. Their "case is set," and Gremio and Tranio "come into court," or, rather, to get a decision on their courting; when suddenly old Baptista—who, we take it, is Shakespeare's ideal of a doting father, especially so far as Bianca is concerned—announces that he will bestow his favorite daughter upon that suitor who will "do the best by her in the way of a marriage settlement." In other words, loving her with all the strength of his paternal soul, he will put her up at auction between them, and knock her down to the highest bidder! The scene which ensues is somewhat like an auction, and something after the manner of a game of bluff, as it turns out. The ardent old Gremio puts up a house within the city, richly furnished with plate and gold, basins and ewers, Tyrian tapestries, costly apparel, (see inventory in the text,) ivory coffers, cypress chests, tents and canopies; ending up through an auctioneer's catalogue of household-stuff, with "pewter and brass," and all things belonging to house or housekeeping, besides one hundred milch-cows at his farm, six score fat oxen standing in the stall, and other farm produce "too numerous to mention." Lucky it was for old Signor Gremio that the city assessor of Padua wasn't by just then, or next year some accomodating

Jew of Padua would have been carrying his tax certificates for him at twenty-five per cent! Then, up comes young Tranio, with much less "pewter," but a good deal more "brass," and—unabashed by the inventory of goods and chattels—is ready to bluff old Gremio with three or four houses in rich Pisa, each one better than his, besides 2,000 ducats by the year of fruitful land, for Bianca's jointure. Somewhat staggered by this, old Gremio brings up his reserves in the shape of "an argosy now lying in Marseilles road,"—and thinks he has choked his young rival with the argosy. Again Tranio "comes up smiling," "sees him" and "goes him better" by three argosies, two galliasses, and twelve tight gallies! Not owning a stiver in the world, this young scoundrel is ready to pledge an unlimited number of gallies, while in truth, instead of any of the gallies coming to him, there was far more probability of his being sent to the galleys! Not satisfied even with this offer, he "caps the climax" with the "bold bluff" that he will double his rival's next offer whatever it may be. It was a case of "doubles and quits," for honest old Gremio sorrowfully confesses that having given all, there's no sense in promising more, and that his rival has won the game.

Scene two, act three, opens on the appointed wedding day of Katharine the Shrew; the priest and friends assembled while the gay bridegroom has as yet failed to "put in an appearance." Already, for some unknown reason, Katharine has become as anxious now to be wed as she was formerly averse. If she scolds now, it is at the delay of the groom, and is about to depart weeping, when it is announced that Petruchio is coming,—but surely in such questionable guise as no woman of spirit would allow a suitor on their wedding day! Tagged out in rags, on a scurvy, bobtailed nag, with the rag, tag and bobtail of the city shouting at the heels of a steed that is caparisoned as outrageously as his master—he appears, at length, and claims his bride's attendance at the church for the ceremony. Although the friends remonstrate, the bride utters never a word of protest. Of his brutal conduct at the church, the like was never known at the wedding of any of the first society of Padua,—and indeed, would scarce be credited had not Shakespeare chronicled it! And still, "the poor craven" bride "said never a word." Beshrew me this is no shrew, or methinks if there ever were anything of the shrew about her she is sufficiently tamed, and it is unnecessary to prolong the ordeal! So thought not the Master, or the play would have been finished in three acts, instead of the orthodox five. Only when required by her tyrant husband to start off immediately for his home, without partaking of her own bridal cheer, did she rebel; and reasonably so, for when she smelled the savory fumes of the wedding feast she had "no stomach for the journey,"—but 'twas only for a moment, for, storming at her as "his property, his goods, his chattels, his horse, his ox, his ass, his anything," he compels her to leave her father's house, in the height of hunger and distress—on his spavined, knock-kneed old horse, and on an empty stomach!

At his country house, when they reach it, he continues this pleasant play of the bedlamite ruffian, the details of which proceedings, though of infinite humor, need not here be set down. Too thoroughly frightened and cowed to apply to the Probate Court for a jury to investigate his sanity, or to send to Padua for a lawyer and make application for a divorce, Katharine submits through two long acts of the play, to all his cruel and senseless vagaries! For a time he suffers her neither to eat nor sleep, and this enforced condition of starvation and insomnia he considers the best joke of the season, terming it "killing her with kindness." Compared with *this* mercenary and brutal wretch, "Peter the Pumpkin-Eater," who merely sequestered his wife for awhile, to cure her of gadding termagancy, by squeezing her into a pumpkin-shell with the chance of eating her way out— this old hero of Mother Goose was an uxorious husband, and a courteous, high-toned gentleman.

The triumph of amiability is completed,— as well as the perfection of probability attained —when at last, on their way back to Padua for a visit to Old Baptista, Katharine acknowledges that two o'clock is seven,—that the sun is the moon,—that old man Vincentio is a blushing young virgin, and addresses him accordingly! If complete asinine stultification be the highest wisdom in woman,—then was Petruchio's conquest of Katharine greater by far than the capture of a city! The abject surrender of her will, her intellect, her conscience, Katharine thus celebrates:

> 'The sun, it is not when you say it is not,
> And the moon changes even as your mind.
> What you will have it named, e'en that it is,
> And so it shall be for Katharine."

Confessing that to her bedazzled eyes, everything looks green, no wonder now that she is ready to swear that the moon is made of green cheese if he so list,—a state of mind most desirable in every well trained and disciplined wife of the Shakespearian model!

And lastly,—at a happy feast at old Baptista's house —when all the jarring elements of the family are well reconciled; when Lucentio, after some little tediousness of trouble, has won the lovely Bianca for his bride; when

Hortensio has solaced his disappointment with a widow fair, fat, and forty, and flush of florins; while the laugh and jest go round, it is only blunt Katharine the Shrew that keeps within the bounds of womanly modesty, not to say, decency!

Then when the wives have left the table and room, a wager has sprung up between Petruchio, Lucentio and Hortensio, as to whose wife is kept under the best subjection—a bet to be decided in favor of him whose spouse shall most promptly return to the dining-hall upon a single summons from her husband. To the surprise of the others' husbands, Katharine is the only one to obey, and meekly, upon further order from her liege lord and master, she lectures the rest as to the proper duty of a wife. In all the ages of womanly debasement, never, perhaps, was the justified slavery of a sex put more completely into words:

> "Thy husband is thy Lord—thy life, thy keeper;
> Thy head, thy sovereign. * * * * * *
> Such duty as the subject owes the Prince,
> Even such a woman oweth to her husband. * * *
> I am ashamed that women are so simple
> To offer war where they should kneel for peace;
> Or seek for rule, supremacy or sway,
> When they are bound to serve, love and obey! * *
> Then vail your stomachs—for it is no boot;
> And place your hands below your husband's foot!
> In token of which duty, if he please
> My hand is ready, may it do him ease!

If the yoke of woman's bondage pressed heavily on her neck for centuries after; if the day of her emancipation, the recognition of her god-given equality of rights, was long in dawning,—retarded by the thousand influences of tradition in laws and in literature,—good reason has she *not* to bless the name of William Shakespeare!

THE EMIGRANT OF 'FIFTY-NINE.

Far out upon the Western Plains,—
In suit of "butternut" and "jeans,"
A man did weary journey wend,
With wagon built at old South Bend,
Containing "grub" and mining tools,
Drawn by two melancholy mules;
And a "yaller dog" did lazily lag on
—No wag in his tail,—at the tail of the wagon.

Unknown my hero's race and name,
I scarcely know from whence he came;
He favored "Posey,"—but more like
He was the flower— and pride—of "Pike."

"What sought he thus afar?"—you say—
"Bright jewels of the mine?"—well, yea!
He was, as you perchance divine,
The "Emigrant of 'Fifty-Nine."

Where Rocky Mountains' streamlets rolled
To sandy Plains their sands of gold,
His mind was fixed his course to strike
Toward his namesake Peak of Pike.

As he through Kansas towns had passed,
They plied him questions thick and fast,
'Till bothered, flouted, jeered and jibed,
He had his "motto" plain inscribed
In letters black, which sprawled all over
His dirty-yellow wagon cover.

What was his motto, do you think?—
'Twas not, "Be merry, eat and drink,"—
Not, "Labor omnia vincit,"—nor
The Alpine boy's "Excelsior!"—
Not, "On to glory's highest star,"
"Ad astra per aspera;"—
Nor dollar's scroll,—"In God we trust:"—
'Twas simply this:—"Pike's Peak, or Bust!"
* * * * * * *

Along the bottoms, rich and raw,
That lined the lonely Arkan-saw,—
And by the sand-hills, one by one,
He traveled on—he traveled on!

Ever along his weary way
He saw the prairie-dogs at play;—
In countless throngs that come and go
From plains to water-gullies low,
He saw the black-browed buffalo:—

Faint shadows on th' horizon dim,
Fleet antelope did sink—and swim,
And fade into "the dying day,"—
While o'er the Plains—"and far away"

And ever toward the setting sun,
He traveled on—he traveled on:—
And to the purple mountain's rim,
That "yaller-dog"—it followed him!

* * * * * *

At length he came where shadows cool
Drape thy clear stream, Fontaine qui Bouille!
Where porphyritic columns odd
Strange semblance take of heathen god:
Where ice-cold fountains seethe and boil,
He rested from his journey's toil;—
And wasted vigor would recruit
With bubbling draughts from "Iron Ute."

Hail "Iron Ute"—Hygeia's shrine!
No fount in all the world like thine!
Old Hudibras should be revised,
In light of science modernized,
And—tribute to thy glorious spring
Ferruginous—we grateful sing:—
"What healthy blessings now environ
The man that gulpeth down cold iron!"

Our Pike went round from spring to spring,
And found delight in everything:
With "Sody-Springs" that foam and fizz
He made "the peartest" biscuit "riz,"—
Not clammy like his usual prog,—
No longer "yaller" as his dog.

* * * * * * *

Again he toiled, a'most a week,
And gained the summit of Pike's Peak.
Who shall describe that glorious view?
Not I,—you'll have to climb there too!

Northward, he sees the ranges roll
Their snowy crests toward frozen Pole;—
Westward and south, in grandeur wild,
Rise mountains back of mountains piled!

With summits capped by wintry snows,
Their ranges verdant Parks enclose,
While splintered peaks jut high in air:—
On this side—mountains everywhere!

* * * * * *

Unknown and cold such Future lies:—
Backward he gladly turns his eyes,
And seeks to measure once again
Two hundred miles of level Plain.

What prospect doth he first descry?
What level meets his homesick eye?
Of all the plain his track had cross'd
Just that which had been longest lost!

The farthest off in vision's scope
Which sinks not 'neath th' round world's slope!
Below him spread unnoted, vast,
The weary wastes he since had passed,—

But in th' horizon's pearl-gray sea,
The waves of earth's convexity
As billows rolling mountain high,
Dash upward 'gainst the eastern sky!

* * * * * *

It is not only in our dreams
The farthest off the nearest seems;
'Tis ever Youth's horizon lies
The nearest to our fading eyes,
And brighter far its distant scene
Than all Life's Plains that lie between!

ON GLACIER POINT, YOSEMITE.

OUT FROM a forest dark with lofty pines
                and cedars straight and tall,
We rode into the open sunlight,
                and trod Yosemite's wall.

Of the world that then broke on our vision,
                so wonderful and new;—
Of that surging chaos of granite
                that pierced the ether blue;—
Of its pinnacles, domes, and towers,
                in such mad proportions given,
They seem "to play fantastic tricks
                in the very face of heaven,"—
Of Merced and Yosemite,
                twin children of the skies
That catch the rain-drops at their source
                —the clouds in which they rise,
Then slip and glide o'er polished ways,
                —then shoot adown the walls,
At single leap ten times the depth
                of famed Niagara Falls:—
Of El Capitan, and Brothers Three,
                Cathedral, Sentinel;
The Lake that gives all back again;
                —it is not mine to tell;—
But that I stood on Glacier Point,
                and what to me befel!

ON GLACIER POINT, YOSEMITE.

The Point—a level granite cliff
        projecting o'er th' abyss:—
On hands and knees I ventured out
        and peer'd o'er edge of this:
On to the far horizon's verge—
        peaks piled with wintry snow!
Beneath me a verdurous valley
        —four thousand feet below!
Four thousand feet as th' plummet drops,
        four thousand feet down straight;
A perfect garden of Eden below;
        —and this the open gate!

And as I gazed and gazed,
        a fascination grew;
I longed for that green valley—
        it seemed so easy, too!
Long hours of toil to climb so high,
        —a minute would take me down!
No thought of a battered body,—
        no thought of a broken crown;
Sam Patch had no such immortal chance
        when he jump'd at Roch'ster town!
From foot to head, in reason's spite,
        th' electric impulse thrill'd me
To spin me down that dizzy height
        though sure as fate it killed me!

I didn't do it! I got off that rock,
        —I daredn't stay any longer:—
Do I falsify?—you go and try—
        and see if your nerves are stronger!

THE DOUBLE BLESSING.

(An incident in the early history of the "Old and New" Club.)

CHEMIST PROF. PATRICK—may he ne'er grow less
But rather double into blessedness,—
Discoursed the Club, one night, a learned thesis
On Bastiat's hobby of Biogenesis;
And fairly proved, to all within the room,
That wondrous life from turnip-tops may bloom.

Discussion o'er, we sought our usual haunt,
To sup with him at Johnson's restaurant.
Now why, when "Old" and "New" alike had gone,
Prof. Patrick lingered in the rear, alone,
We never knew, but if I guess aright,
To close the club-room, and put out the light.

How sad it is—as we so often find—
That out of sight's the same as out of mind!
How strange it was that just as soon as ever
We saw that feast we clean forgot its giver!

That feast appointed for the stroke of ten:—
Minutes are hours, sometimes to hungry men!
Short grace was said, and then, oh, grievous sin!
Dwight gave the signal, and we all "pitched in!"

But when the host appeared at last, tho' late,—
Each guest sat silent—and looked on his plate,
And fain to cover up his fault of breeding,
Tried hard to make believe he'd not been feeding!

Exceeding peace had made Prof. Patrick bold,—
He kindly glanced around the "New" and "Old,"
And clearly spake, as was his proper task,
"Prof. Robinson, will you the blessing ask!"

Quoth Dwight,—"Professor Patrick! nay, not so!
The blessing's asked a good long while ago."—
Again spake Patrick,—but he spake more low,
Yet cheerily still,—and said: "I pray you then
Since I, though host, have been forestalled by ten,
Set me down one who will my duty do
Upon the food already blessed by you:—
For sure the poet sang, long ere to-night,
That blessings brighten as they take their flight!"

Then did Dwight yield, and Robinson subside;—
While down the line, as by the waiters plied,
Some answered "raw," and others some said "fried:"
But two, in spite of Patrick's courteous healing,
Thought "stewed" alone expressed their proper feeling!

But all that feast, the truth stood fair confessed,
Gracious or graceless,—blessed or unblessed,—
Each man undoubted did "his level best,"—
But lo! Prof. Patrick's plate led all the rest!

www.ingramcontent.com/pod-product-compliance
Lightning Source LLC
Chambersburg PA
CBHW030017240426

43672CB00007B/994